Listen!

New Discoveries about the Afterlife

Scientific Research on Contact with the Invisible. Experiences of Instrumental TransCommunication (ITC)

Anna Maria Wauters and Hans Otto König

For Katharina and Markus

4

Greater Reality Publications
A Division of the Afterlife Research and Development Institute, Inc.
23 Payne Place
Normal, IL 61761
information@afterlifeinstitute.org
800 690-4232
www.greaterreality.com

Cover by R. Craig Hogan

Translated from the original French book,

Anna Maria Wauters

NOUVELLES DÉCOUVERTES SUR L'AU-DELÀ

by Evelyn Meuren,

with editing by R. Craig Hogan

The translation and publication were performed as a service of the Afterlife Research and Education Institute, Inc., to promote Hans Otto König's important research.

Contents

Preface

We live in a remarkable period in humankind's history. About 70,000 years ago, a cognitive revolution propelled humankind into advancing in the ability to solve problems and think creatively. Great milestones in human advancement followed. Around 12,000 years ago, humans ceased the nomadic hunter-gatherer lifestyle by growing crops, tending livestock, and establishing communities. By 8,000 years ago, modern humans used symbols to represent words and concepts. Great advancements followed in waves: the Renaissance of the 14th to 17th centuries; the scientific revolution in the 16th through 19th centuries; the industrial revolution of the 18th to 19th centuries; and the technical revolution in the latter part of the 19th century. Quantum mechanics caused science to reconceptualize the nature of reality in the first third of the 20th century, and the information/telecommunications/digital revolution began in the last half of the 20th century. Every one of these milestones was a great leap forward in humankind's development.

Today, we are witnessing the development of the next great leap for humankind. Through the pioneering work of Hans Otto König, humankind is communicating with higher-order Spirit Beings from other worlds. These beings are concerned with humankind's welfare and are diligently communicating with König to open the channels for communication that will eventually result in regular dialogues to help humankind.

The reason these advancements could not have occurred in earlier times is that the communication media have been available for only a few decades, made possible because of the harnessing of electricity in the 19th century. Samuel Morse used electricity to power the telegraph in 1837. Alexander Graham Bell transmitted voice through a telephone in 1876, and in 1894, Oliver Lodge used electricity to manipulate radio waves to convey information. Marconi transmitted voice using radio waves over a longer distance in 1897 and continued to develop wireless signals into the early 20th century. The first television was built in the United States by V. K. Zworykin in 1929. In the middle of the 20th century, Walter Brattain, John Bardeen, and William Shockley invented the transistor, and in 1958, Jack Kilby and Robert Noyce invited the integrated circuit. In 1962, the Telstar satellite relayed TV signals between Europe and the United States.

As a result of these developments, sophisticated electronic communication devices are now available for the first time in humankind's history. Realizing this, teams of dedicated Spirit Beings in other realms resolved to help humankind learn to communicate with their worlds using devices previously unavailable to humankind. The Spirit Beings have chosen individuals on the Earth plane to team with them in this important work. The most successful pioneer on the earth plane is Hans Otto König, who has devoted 45 years of his life to technical research in his laboratory, continually developing new, more advanced devices to communicate with Spirit Beings. The findings from his research will be of great interest to the scientific world in general and basic researchers in particular, now and in the years to come.

König detected the voices of Spirit Beings seeking to develop enhanced communication with humankind on the earth plane for the first time in 1974. Over the next 45 years, he passionately studied the acoustic phenomena producing the voices, extending the reach of the technical apparatuses with which he worked far beyond what was understood to be possible; he was able to improve the signal strength and communicate with the Spirit Beings. They were anxious to share knowledge, so they participated in improving the technical apparatuses König was developing. Today, he regularly makes contact with the Spirit Beings over the connections called "contact bridges" and is sharing what he has learned with other researchers and humankind.

He has learned from his communications that the entities living in the other realms are at higher levels of being: the 5th through 7th levels. These Spirit Beings have been in contact with König throughout his research.

They continue to transmit messages to us that they want humankind to hear:

- We are eternal beings. There is no death as an end to the individual spirit. "Nobody ceases to exist; every life is forever," they say.

- The universe must be penetrated and directed by the forces of the Spirit. That is why man is necessary in the Divine's eternal plan.

- The Spirit Beings have a great concern for what is happening on the Earth plane. They send the power of Love to help us.

- They have told König that the time is ripe for an expansion of consciousness.

- The connections will become stronger. The contact bridge will grow in power and capability.

- Many people who lived on the Earth plane want to connect with us now, especially loved ones.

- They assert that extraterrestrial life, other life forms, will contact us.

- Good connections require inner preparation and a spiritual focus. Not everyone can make good contacts with us by special, more advanced devices. Stature and spiritual maturity are the bases.

- "All is one." We are connected to them in love. We are all a unity of life, even animals and plants. The meaning and value of life are based in love and the union with all that is alive.

- The Spirit Beings know our thoughts. They report that they are "watching" us.

- If we want to change the World, they say we as individuals must change our own worlds, within each of us. They assert that spirit overcomes matter.

These profound insights have been given to König by the Spirit Beings for the benefit of humankind. He is now anxious to make their teachings available so all people can benefit from them, fulfilling the Spirit Beings' purpose in connecting with us on the Earth plane. This book contains the essence of these teachings.

R. Craig Hogan, Ph.D.
President, Afterlife Research and Education Institute, Inc.

Hans Otto König, 1982 in his lab

"Everything is born of Spirit; everything is Spirit. The basic substance of all that is, is of a spiritual nature. If one day we are able to realize that we are all Spirit, having a physical body on the Earth plane only as a simple shape for expression, then we will extend our knowledge beyond the physical world. Our minds can communicate among ourselves, whether connected to a physical body or not. All is one! Those realms we call the "down here" and the "beyond" are not separated. It is the same world, but with different material qualities. And "beyond" is not anywhere up there, as is commonly envisioned. Although we cannot perceive the "beyond" with our five senses, it is here, where we are living now. It is time for humanity to expand its worldview, which is unfortunately still based on the familiar five senses. It is evident that vibration frequencies are passing through walls and through houses, although neither seen nor heard. We are surrounded by information that we are unable to detect with our five senses. Twenty-first-century science remains confined within the boundaries of outmoded thought; it is now important to extend our mindsets. The permanence of personal consciousness for me is a fact with too many demonstrated proofs to be ignored.

Parallel worlds do exist. I am a firm believer in that. If that were not true, none of my technical devices would have given the amazing, spellbinding results they have given. In the dimensions beyond our five senses, "information fields" do exist and can be detected, if detecting them is aided by adequate material means."

Hans Otto König

Prologue

"Tell all humans that we are alive!"

This knowledge, the certainty that life continues to exist beyond physical death, would change our way of life fundamentally and modify our attitude toward life radically. The metaphysical questioning about this great unknown has concerned humankind for ages. But in the modern world, where we confine our conception of reality to matter by refusing to acknowledge the reality of the invisible mind, a split has developed between science and spirituality, between knowledge and belief, between physics and metaphysics. Official science is unanimous in holding these dogmas: Our consciousness will die at the same moment our brain dies. The conception of life after death is only a matter of faith.

In today's world, however, dominated by technology, an acoustic phenomenon that disturbs this official science dogma has emerged: The voices of deceased people living in another reality that is a parallel world to ours have been recorded. These voices tell us the people living in the parallel world want to get in touch with the human world and transmit information to it.

The German high-frequency, electroacoustic physicist Hans Otto König detected these voices for the first time in 1974. With his training in high frequencies and technical background in electroacoustics, König was fascinated by the voices, but convinced they were created only by his own subconscious mind. Over the next 40 years, he passionately studied this exciting acoustic phenomenon, resulting in discoveries totally in contrast to what had been his knowledge of common physics principles.

"LISTEN!" The word often punctuated these live connections with another world during his research. And this open-minded man listened. He was raptly attentive to their messages, asking inquisitive questions:

What is the nature of this completely new and inexplicable phenomenon?

How does it work?

 Why does it exist?

During innumerable trials, König listened with his whole heart and soul, analyzing what came to him through his devices, which are called by the invisible beings speaking to him "contact bridges." The entities on the other life planes helped him develop new devices that allowed him to experience clear connections to Worlds Unseen, making thousands of messages audible to humankind and developing several new devices with the help of those on other life planes, establishing what is called a "contact bridge" that allowed him to get clear connections to the Worlds Unseen.

This technical contact bridge allows us humans to make the mental structures of the Spirit World audible for the human ear. They give us a distinct material contact, allowing people to gain more knowledge of THEIR spiritual nature.

The many technical discoveries and objectively measurable results and the acoustically audible statements from the Spirit World convey a truth that reduces the gap between science and faith and lifts briefly the veil between this world and the next. Through these phenomena, they are demonstrating that all is one. Serious study of these voices, however, presents orthodox science with a significant challenge that is forcing a revolution in their paradigms and worldview.

Considering the question, "What happens after physical death," ultimately, while it is a question about death, it is especially a question about life: What is life, man, the living being in the visible and the invisible region, en masse or in a microcosm, in this world and in the next?

The voices are like songs—sweet melodies of being—transmissions unlike the public dialogues occurring all around. The voices are golden filaments woven between the levels of conscious-ness, bridging the gap between science and belief.

For the first time, Hans Otto König is speaking out in this book, talking about his experiences, his discoveries, and his re-search. Technically, the experiences are one of a kind in the world. Not much has been said and written about this researcher, but he has persevered. He has continued to create new contact bridges, even though his unparalleled research is hardly known. These pages are exclusively devoted to Hans Otto König's research, even though there are many ITC researchers in the world with more fame and recognition, and there are a great many other publications dealing with the topic. Nevertheless, it is a rare and unique experi-ence to accompany closely an "Archaeologist of the Spirit World," walking for many years in both this world and the next. This testi-mony, rising out of long conversations between the researcher and the Spirit World that pursued him from childhood through to the period of his recent developments, allows him to confirm for us with certainty that the personal mind of man continues to exist be-yond death, in worlds parallel to our own.

"Listen!" they said to him.

"Many people live with the thought of the absolute loss of a loved one. But this is not a loss in the true reality. Nobody ceases to exist; every life is forever!"

Note : The bolded statements in the book are the messages from the Spirit Word

First Part

The Path of a Singular Researcher

*On the cosmic scale, only the fantastic
has a chance of being true.*

~ Pierre Teilhard de Chardin

Introduction

Childhood. You read us fairy tales. We listen big-eyed, imagining Snow White, Bluebeard, Cinderella, The Little Mermaid, Puss in Boots and many other fairy tale-characters. And in fairy tales it is quite normal that objects can fly, animals can speak, and all nature is a living being around us. A child's mind is convinced that its best companion, his pet, must be his guardian angel, protecting the child always and everywhere and that the woods are inhabited by monsters, dwarfs or fairies. Children are discouraged by adults from believing all these beautiful stories, explaining that fairies and angels exist only in our imaginations, that reality has quite a different face and that the dead are obviously stone dead; we can keep them alive only in our hearts. But what do they know about our true nature? Is it not revealed by these wonderful stories? Don't forget, those initiatory narrations bind man to his homeland, to his inner reflection about the union of Earth and Heaven! Why do people realize so little, or even not at all, the many phenomena that occur at the borderline between "the natural" and "the supernatural"? In addition, "Could it be that they don't want to consider that the paranormal is more normal because the paranormal bothers us enormously?" as R. Chauvin so aptly stated. So the one who takes the trouble to listen to the whispers of the Spirit entities and explore the Spirit World will realize that what passes for fantasy does exist indeed, that matter is transformed by "the mentally visualized magic wand of thought," and the crystals capture and transmit much information. He will sense the dead giving him signs from another realm and understand that consciousness doesn't stop at the limits of his brain, but opens up the unlimited Life of Spirit. That's what happened to H. O. König, a German researcher, who could be a character in a story titled The Solitary King — yes, König means King in German — listening to other Worlds.

Listening to this man speak about his life's journey, he usually appears so calm and reserved about his private life that you're caught unaware by the wonderful insights he realized at the very

21

beginning while he surmounted life's inevitable difficulties and typical trials, that he narrates in beautiful childhood stories that were whispered directly to his soul about the imperceptible life in the World Unseen. We know that learning about mind and matter can be a difficult, laborious and treacherous narrow path rife with obstacles. Many times the experimenter was at the point of giving up the adventure of his singular research. But on his way he was accompanied by Spirit; he was even chosen by them. His wife, Margaret König, a well-known hypnoanalyst, was, for many years, the only one who shared with him all his doubts, his reflections, his sorrows and his immense joys. She supported him in any situation, often inspired by his strong character and independent spirit. The major part of this important research, that was at the same time a quest for knowledge—even though the researcher never uses those terms—was made in fruitful solitude. So it was particularly important to have a heart at his side he could absolutely trust, to face with him the surrounding world of matter that often reacted with hostility to this kind of research. His wife was close to him during the transformations and many paranormal phenomena, which the scientist always kept to himself, unable to explain them. She witnessed the awakening of his subtle sense of his psychic faculties, as well as his ability to induce a trance. She alone walked the path of interiority and knowledge with him. Together they met with many grieving people, people with incurable disease and anyone who simply had questions about life, death, and the other world. If Margaret hadn't become so ill, she would have written a book about her husband's work. König often said, "Without the support of Margaret, my research could never have been developed in this way." Their life was a fight, a journey, and a beautiful adventure of consciousness.

Reflecting on the life story of this explorer of the World Unseen, you wonder whether certain people are born with a particular destiny to fulfill during their life on the Earth plane. Because we can recognize in this young boy great predispositions for courageous pioneering, he was to accomplish later in his life. Studying this child, we get a little better understanding of why he has become precisely that man with that open-minded scientific spirit, working for large German companies, taking seriously the barely audible whispers from the nowhere. How he had patience, perseverance, and courage—even if someone would like to call it mania—to do such extensive research on the Spirit World, risking laying himself open to ridicule by traditional scientists who still seem so often to turn away from the telescope to observe again with their naked eyes "the reality of heaven." Hans Otto König's great success shows us once again, how much desire, thirst for knowledge, patience, perseverance, interiority, willingness, pureness and the strenuous effort of a curious child can guide us toward new horizons and new tomorrows.

The Time of Childhood and the Beginning of the Research

Hans Otto König was born in Düsseldorf-Ratingen, Germany, North-Rhine Westphalia, in the middle of the 20th century. He had a warm-hearted mother who was an opera singer and talented actress, and a strict father who was an engineer, an excellent chess player, and violinist, owning the only radio-and-television shop in the small bourgeois district, not far from the banks of the Rhine.

But in his mother's womb, life and death would form an inseparable pair. Indeed, while the little Hans Otto was about to come into the world, his little sister Marie suddenly passed away from a dazzling disease at the age of 5. Shortly before her death, she said to her mother, "I have a little brother. I'll not meet him, but you shall call him Hans Otto, won't you Mom?" Her mother didn't understand little Marie and laughed at the words passing the little child's lips, because her beloved daughter was in excellent health. "Certainly you will meet him, your little brother!" However, the young woman was going to experience the death of her little girl while new life was growing in her. Death and sorrow would always be present during little Hans Otto's childhood, as well as bombings, shootings and all the cruelties of World War II.

During his childhood, one person in particular would play a central role in Hans Otto's life: his beloved grandfather! This man by trade was a master carpenter and was a cultured autodidact with an alert mind. He worked part-time as a stage director at the theater in Düsseldorf. He was a righteous man with a big heart and extraordinarily strong character. This independent-minded man was devoted to his grandson Hans Otto, and during the long winter nights, he loved to tell stories by candlelight about the life of "Schinder-Hannes," a prominent rebel who radically refused to bow to the official authority and pledged to rectify the injustices done to the poor. He taught Hans Otto how to use a shooting knife, gardening, and manual labor, but above all, he tried to forge independence in the child's mind, introducing him to life in long conversations, reflecting on all the experiences he had during his own long life. He even guided his little grandson to find his inner self, assimilating impressions deeply to create his personal mentality. To set examples for little Hans Otto, time and again he talked about his own experiences, especially those he had during World War I and World War II.

Hans Otto's grandfather also invited the boy to accompany him through stories on all his many travels around the world. Hans Otto still remembers the stories about China, where his grandfather fought as a German soldier in 1901. More than once, he stared death in the face, but he had never killed anyone. In China, he learned the martial arts, learned about Buddhism and especially learned how to meditate. Each day, this German Catholic man, at a moment's notice, would retire for 20 minutes to find his inner silence. In such moments, everyone had to leave him alone. In his silence, he learned to compose himself. He and his wife, Hannchen, were a very devoted couple, happy to share the rough and the smooth experiences in life. Hans Otto's grandmother often said, smiling at the boy, "Do not listen too much! Your grandfather is something special, a little bit crazy!" But the little grandson loved to listen. Free and open-minded, the grandfather never shrank from any question posed by his grandson, and never hesitated to discuss with the boy all imaginable topics.

At the end of his life, his grandfather promised Hans Otto to accompany him all his life. Only many years later did the boy really understand the meaning and depth of these words. He later realized that his grandfather had initiated the pioneering attitude to develop Instrumental TransCommunication.

The Power of the Intimate and Profound

This alert-minded ancestor, having eyes like a hawk, ventured to explain to the child, the power in a deep and intimate relationship, freely chosen, with another being. You cannot create this link; it is there or it is not. But we can cultivate and nourish it, or simply let it wither. This is a link where honesty with oneself and the other is a divine command. In this rare relationship, neither would hesitate to give his life for the other. This is a gift from heaven, precious to them, a grace to be held dear, deserving the utmost care.

The child, revolting against any kind of authority, could always count on his grandpa, who, despite his advanced age, remained young at heart and in mind. For example, it was he who stood against all in support of Hans Otto at an event that had taken place at the Catholic Church in his little town. Hans Otto, age 13, was an altar server in the church. In those days, religion forbade a layman to touch the monstrance; you might drop dead if you dared to. Hans Otto didn't think so and thought he would show them it was nonsense. As proof, one day he took the monstrance in his hand, remarking, "Look, there is absolutely nothing going on." The other boys, having nothing better to do, quickly spread the word about the incident. My God, what sacrilege! A bishop of Cologne had to be summoned to bless the monstrance of the church of Ratingen again, and the little blasphemer was accused of having made a pact with the devil.

After that, when people saw Hans Otto, many crossed to the other side of the street, refused to talk to him, would not sit next to him and would not serve him in the store. But his grandfather just laughed about the whole matter. "You only wanted to realize it for yourself; that is more important than anything else."

The Power of an Independent Spirit

One day Hans Otto returned from school and went directly to the workshop to meet his grandpa. Without Hans Otto's saying anything, his grandfather asked him, "Well, what happened with your classmate this morning. Tell me!" Hans Otto was astonished. "How do you know that, Grandpa?"

Such episodes were often repeated, to the point that they convinced the boy that his grandfather could read his thoughts. They had lengthy conversations about the power of thought and telepathy. Very often, his grandfather told him about his premonitions. But then, what is the meaning of the term "Time"? Is it really "the dark enemy, that eats into our lives" that Baudelaire talked about? "Time" was one of the most dominant issues, and it remains so.

Hans Otto today is today the age of his beloved grandfather, reliving other scenes of his childhood. "One day I had beaten some boys up badly at the playground. My grandfather listened to my story and simply asked me, "Why do you fight with your body? Why don't you use the power of your mind?"

I asked him, "But how?"

He said, "First learn to control your mind."

"How do I do that?" I asked him.

"Quite simply. Focus on a candle, first for half an hour, then for one hour."

"One hour! But I will never have the patience to do it."

"The first time it will be a little difficult, but you will learn how to do it. Patience is everything in life, and this time you have to persevere. A whole lifetime has enough time to reserve an hour for yourself."

The next day, I told my grandfather that I had just tried it, but was able to sustain it for only three minutes. He replied, "Great.

Now you know you can do it, so keep on doing it. Let it become a habit!"

I meditated every day, and after some time, I was able to focus so intently on the candle flame that I didn't even realize when someone entered my room or called me. After my grandfather taught me this lesson, we often held long talks about the power of the human mind. My grandfather's probing questions about my sensations and experiences during this mind training brought me to realize what an enormous power our minds hold, that the human mind can use its power to create wonderful and positive things as well as destructive and awful things. Everyone has a purpose. He taught me that no one has the right to impose his personal thoughts on someone else; everyone must have the power to decide for himself, after careful consideration, whether he will accept or reject the thoughts of other people, even at the risk of his life.

The Power of the Inner Eye

Hans Otto's grandfather learned through several war-time experiences that survival can depend upon being able to recognize a person's character. Apparel doesn't matter, social status is irrelevant, and money means nothing. Instead, it is important to know how the person standing in front of you looks inside. This extremely sensitive man wasn't swayed by titles, honors, status, appearance or even the kindness someone displays. He had developed the ability to see inside a person, with a penetrating discernment. He endeavored to give this gift of inner discernment to the young Hans Otto. The ability to see past the physical body into a person's inner being is a remarkable talent with immeasurable benefits. The mindset necessary to be able to exercise this ability to view another person's inner being is developed by training the mind to observe the smallest detail with utmost concentration, directness, and honesty, without concession. During wartime, these terms have been overused, but they still play important roles in our lives, just as the term "Pureness of Heart" does.

Love and Respect for Nature

"Never stop observing people and things. Keep strictly observing beyond what you see at first sight," Hans Otto's grandfather said to his small grandson. Hans Otto often preferred spending hours alone in the woods instead of going to school. He adored nature's thriving, natural growth, ALL of it! He wanted to learn about ALL of it! For long hours, yes, even for days, he observed the finely tuned, structured movements of an army of ants. A fox family came to trust him, even letting him play with their puppies. How angry he became when he saw other children smashing leaves with sticks, trampling flowers, or scaring weak pigeons out of sheer wantonness. He loved to watch a plant budding day by day and peering through his grandpa's microscope at drops of water from a pond or a leaf or simply drawn out of the soil. He loved studying the famous "frogspawn." And on and on. In wonder, he discovered complete microcosms were opening up to him in images not visible to a naked eye. It is an incredible world of discoveries made for someone who wants to explore the universe up close. People who feel no desire to see the world more profoundly will never experience the miracles available in nature, and will never discover all the secret worlds that seem invisible until first seen. Many years later, Hans Otto's encounters with the paranormal voices would give this intense lover of nature answers to some of his questions, enabling him to see clearly that "the essential is often invisible," and that off of a main street are always a great many side streets.

The Love of Music

During the years young Hans Otto was receiving lessons about nature from his grandfather, he regularly wandered into the woods. There, sitting against a tree, he would sense the rhythms of nature, realizing that every location has its own music welcomed by the silence of his inner self. He loved listening to the melodies singing through the trees, the fields, flowers, and winds. The boy took piano lessons and learned on his own to play the accordion.

Born into a musical family, everyone met on Sunday to sing and make music together. It was obvious that they did not make music by reading a score, but from memory. At 17, Hans Otto founded his own group. He never included any musician who could not remember and play the music by ear.

Music remained a fond memory for him. He embraced every style of music with an infinite love. Citing Nietzsche: "Life without music is simply a fatal error, a fatigue, an exile!" Each life event is accompanied by music giving to it a special meaning. It is a mirror like no other life and one's own life.

The Passion for Technology and Physics

Just as he loved music, Hans Otto was passionate about physics and anything related to technology. Although his grades at school were not high in foreign languages, he distinguished himself by being the best student in physics, becoming "cum laude" in the entire region. He adored this field of study and often went to the community library, devouring books on physics. Perhaps he thought he would find in those books the answers to the questions his teacher had been unable to respond to in class.

At age six, he built his first radio, by himself, with only the information he found in books he borrowed from the library. His love for crafting radios never ceased, and at age 18, he even obtained his amateur radio operator license. Often he stayed up all night exchanging messages with people from all over the world.

Hans Otto always wanted to know how things work. He completed all manner of experiments, very often bringing trouble to his family. If people in the town heard something explode, they could be sure that Hans Otto wasn't far away. He loved experimenting with all subjects and objects to see how they would react under certain conditions.

However, more than anything he felt the urge to find out what the inside of things looked like, so he took apart all mechanical devices. A magnificent alarm clock or the new electric locomotive he received at Christmas were all disassembled into hundreds of small coil springs, bolts, screws, and cogs. Fortunately, he always managed to reassemble the devices and make them work again, even if he had to spend weeks doing so.

For Hans Otto, it became a challenging sport to invent something new that was useful to others. At age 16, long before hearing aids were invented, he tried to construct a hearing aid for his grandfather, whose hearing had deteriorated considerably.

So braving the ridicule of passers-by, Hans Otto's grandfather proudly walked in the Ratingen marketplace, his beautiful silver cane in his hand, a grand hearing aid instrument crowned on his ears and a big wooden box on his chest containing a kind of acoustic instrument, all invented and assembled by his little grandson. He proudly boasted that the 14-year-old had set up an illegal connection to the radio network and was listening to the main news from the region (a serious matter for which Hans Otto was heavily admonished by the public authorities). Hans Otto didn't even tell his mother about the connection. Then, one day a police team armed with pistols and machine guns invaded their house to arrest him, mistakenly thinking he was a dangerous spy. "During the war, your son would have been shot on the spot," the commander later told his frightened mother.

The Urge for Knowledge

Like many young people, Hans Otto was particularly eager for knowledge. Everything related to metaphysics was extremely interesting to him, and he sought out the meaning of anything. He would sit for hours on the riverbanks, reflecting upon the meaning of this or that experience, upon activities performed one way or the other. He didn't understand, for example, why his classmates loved chasing a ball, passing it to someone else and feeling such great joy when the ball sailed accurately between two poles. For him, there was no point in doing any of it. He was more likely to be interested in other sports, like judo, through which one could master the body and defend oneself in a dangerous situation.

He frequently meditated about what constitutes man. He often watched people, sometimes provoking them, through carefully planned situations, to learn their reactions to the unpredictable and unexpected. He noted big differences between what people think and what they say! At the age of 10, he wanted to become altar server in the Catholic Church because of his interest in the procedure of a mass. He did not want to be only an observer; he felt he must know what was happening in the religious ritual behind the curtain. Then he wondered, What is a priest's life like? Do his actions match the words he says? What does he really believe? How does he feel about this? How in fact does a mind work?

He began investigating the newest medical discoveries. Young Hans Otto wrote a letter to a famous surgeon, who had just successfully completed his first heart transplant, asking him for more information on the subject, but he never received a response. He devoured the biographies of all the great pioneers and explorers, passionately reading about their discoveries and lives. He didn't stop asking why things were so and what were those things and were those things more illusion than reality?

König often describes the time when he was lying on the meadow in his grandparents' garden watching the clouds and the vastness of heaven. "Where does the world end, I asked myself. With the imagination of a 4-year-old boy, I thought that somewhere there must be a large fence marking the end, and that behind it would be only total darkness. Then I said to myself that one day I would try to make a hole in that fence, and with my flashlight, I would look in to see if there really was nothing hidden in the vast darkness."

Meanwhile, he listened with great interest to religion instruction, learning about the major issues that arose.

But one day, when the priest told the students, that man came to Earth to fulfill God's will, he thought it was odd, replying that he doubted its validity because it would mean that man doesn't have free will. And how could we know God's will for our life? The priest was enraged, striking him in the face; how dare he question the Word of God when the Church and the Bible showed us perfectly the way to follow them!

But Hans Otto continued questioning. How is it that no one has ever seen God? Where can we talk to him? And how did he, the priest, know that the Bible is the Word of God? How was the world created? Where did we come from? Where do we go after death? What is heaven? What is hell? How could the clergy be so sure of their answers?

Hans Otto quite often disrupted the class with questions. They told him he was insolent and provocative. Years later, connecting to the Invisible World through his research in Instrumental Transcommunication (ITC), he continued looking for answers to all his questions by relying on observation, his own experiences and the innumerable messages transmitted by the Spirit World.

The Time after Childhood

At the age of 18, the young man decided to start studying physics, specializing in technical-high-frequencies, the discipline he was most interested in. At the RWTH University of Aachen (Germany), he began his studies, financing them by working in the world of music. Unfortunately, he did not meet the requirement to graduate with a doctoral degree, for personal and family reasons. Maybe this abrupt stop was his good fortune. If he had been as independent and open-minded as he had expected and desired would he have pursued an academic career? On the other hand, maybe his unconventional research would have had even more recognition in a world that reveres titles, had it been produced by "Dr. Hans Otto König" instead of "H. O. König."

Anyway, the young graduate was earning money and began working for several large companies. Living at a time when Germany was undergoing reconstruction and an economic upturn, he could afford to pursue lines of work he liked, and even to impose his personal working conditions on his different employers. He always allied his work with other passions, like music and photography. In 1970, he started his own business. At the same time, he discovered the electronic voices, a gift that grew out of the mind of a curious child who doted on learning electroacoustics and high-frequency technology. The adventure to enter unknown territory was about to begin . . .

The Time of the Research

A question of fundamental ethics and grave doubt

A Fundamental Ethics Question: "Do We Have the Right to Enter New Territory?"

On the day the investigator was nearing the conclusion that the electronic voices must have their origin in "psychic structures" of deceased people, he encountered a huge problem. Was he entitled to research this invisible territory, only separated from ours by the veil of physical death? How could he be sure that contact didn't discommode "the inhabitants of other worlds"? Shouldn't he respect their "eternal rest," as they say? Was it fair to challenge their existence? Was contact desirable between the visible world and the world beyond our sensory perceptions? Why did an ontological difference seem to exist between these two worlds?

All these issues weighed heavily on Hans Otto König. That's why during the recordings he never wanted to call on a specific person living in those other worlds for answers. The first questions he posed were always the same: Do I have the right to contact you? Am I intruding?

And here are the responses he received:

"We need your contacts."
"The contact causes something great."
"Your link-contact is important."
"Contact is of divine vibration."

No entity ever told him that contact was enervating him or hindering him in his new life. The opposite was true. But if the contrary had been the case, H. O. König would have immediately stopped all his research. So he dared to continue. But doubt remained.

Who had answered him? When recording by radio, he didn't know the dialog partner. So how should he interpret these statements? He didn't have the option of finding answers to these questions in others' books. So on what could he base the statements? Communicating with those called "the dead" is a subject many regard as suspicious and unproven. He faced a paradox: if he continued his experiments, he could receive only a response based on reality. So he proceeded with his experimentation, but with all due respect to the other world.

Years later, this inner doubt was removed completely by his invisible partners, with whom he established a trusting relationship. He learned where the information originated. He got to know them through their multiple transmissions. He was even able to verify "their spectral identity." They told him they were talking to him from a sophisticated cosmic level. They often reassured him by saying, for example,

> **"Listen! The assertions of the Christian Bible are false. You don't doubt the dead. Because we are not dead. We can answer your questions."**

> **"Listen! We repeat. We need your contacts like you need ours."**

> **"Listen! Our development is not hindered by your contacts."**

> **"We welcome all. We are pleased that contacts exist. Contacts do exist."**

These were the messages of these highly evolved spiritual entities. But what about all the others? Here's the opinion of the researcher, spoken during a conference:

> Never forget to stop at the borders of the world on the other side. Never try to *pass them* before asking

their permission with all humility and maximum respect. These are the same people granting you access. Don't try to bind somebody to yourself, as long as we still know so little about the "how and why" of the contacts. We must act very carefully. We still know so little. So how can we be completely sure that certain entities do not still feel obligated to answer your call?

I know, for many spirits, that problem doesn't arise, and among the most advanced entities, I'm absolutely sure that we don't annoy them. But all the others? Do they think of it in the same way? There exist so many "Spirit beings" and even many different "information panels." So approach respectfully, posing your question. Please don't quiz someone about everything, particularly the special one who comes to visit you every day. Certainly that wouldn't be the reason for doing the ITC. It could even be harmful, either for yourself or for your invisible partner. Let it unfold naturally, totally naturally! If someone has something important to convey to you, be assured that he will come, all by himself. Until then, stay outside the front door, and only then pose your question, very simply, asking if anyone has a message for you. And if not? No problem. Please don't be sad! If they send a message to you, be thankful, but then carry on your earthly life, in freedom and independence, footloose and fancy-free. Only minds that free themselves from physical bodies live among us, in a world like ours, even now, after having completed the battle against matter. They're living on different life planes, having new experiences and various interests different from ours.

Don't use them to satisfy your emotional needs and don't employ them as a detective agency!

As long as we still know so little, those who deal with ITC have an enormous responsibility to the presences in the Invisible World. Please don't make the same mistakes already committed by those who discovered new lands on our planet. Don't approach Spirit with the attitude of a conqueror, a missionary, or a beggar, but with the greatest respect for their borders, dignifying them as "Spirit beings, living in those worlds away.

The most advanced minds, for their part, observe an experimenter for a long time before contacting him. They're in charge, if and when you merit it."

The Many Attacks

In his lifetime, a seeker of knowledge will encounter what Western traditions of Spirituality often call "the battle between light and darkness." Either you'll win the day or you'll be attacked by evil forces. These forces didn't spare the researcher and his family, assailing them consistently through the years.

In everyday life, the most exhausting chicanery was often perpetrated by colleagues: history repeats itself. For centuries it's always been the same old story of power and jealousy. For 15 years, he and his wife had to deal with a great number of complaints reported to the police. They even had to face four big lawsuits, all incurred somewhere else. Each time, the researcher was charged with fraud. One day he was denounced for allegedly deceiving millions of people on a TV show where he had conducted a live experiment with nearly four million viewers. Another time, his devices supposedly caused disease. And on and on. There were even numerous threats to his life requiring a bodyguard or his family was terrorized.

One day, he dared to ask if he could present proof to support the authenticity of certain long and short messages received by the supposedly paranormal wire. He was shunned by many people and became the butt of criticism in the ITC World.

Still other problems were heading his way. This time they came from the unseen world when he introduced a delicate topic that evoked fear in many people: Evil Forces came from invisible levels! For many years, the experimenter noticed, every time he was about to take a big step forward in his research, unseen beings in the background strove to counteract it. For example, some of these creatures deceived him while pretending to be part of the "Zentrale." But the researcher, knowing well that scene, realized immediately that was quite untrue! These false ghosts prophesied that soon he would die of a heart attack, just like his mother. Negative forces often tried to provoke "dark hours" in König's life, and in some way or other, they continually caused confusion and angst in the family. After this storm and stress period finally ended, the real friends of the explorer reappeared, stating, **"We surround Hans König with a group of protective forces."** In 2006, his invisible friends let him know the same, but in other words: **"The mantle of our love will give you protection."**

Given all these adversities, the researcher reached the limits of his endurance while gaining new energy and fortitude, and an understanding that polarity between positive and negative forces is not abolished after passing through the gates of other realms. Everyone who intends to follow his inner path will face not only opponents incarnated in a physical body, but also non-incarnated forces. They want to impede the progress and new discoveries that enlighten.

The researcher assures, "It is very important not to be swayed by their intrigues and consider them insignificant." To succeed, it's not enough to know only one's strengths, but also one's weaknesses. You must sharpen your senses and exercise discernment to make the right choice.

A Life in the Service of Research

H. O. König's research wasn't for pay, unlike that of other scientists. He didn't live on his research. It was not an easy task for him to find money for expensive investments for his laboratory and to make a living. Despite this, he still tried to be financially independent so he could conduct investigations in complete freedom, without restrictions, only for the sake of science. This naturally affected the family's material condition.

A little revealing episode follows. Coincidentally, just at the moment the couple was about to buy a beautiful home, the experimenter's research reached a critical point where he needed significant funds to continue developing the Multi-Oscillation System. If he couldn't find the money, all his research would have to stop. He needed a huge investment, but with no certainty of success. Would the new system work better, giving him access to new learning? He didn't know how to decide. His wife gave him the answer: "Well, so we won't buy the house. We'll use the money for your research!"

The couple had to make many sacrifices, both financial and social. It was remarkable that the couple did not succumb to certain temptations of the world, like extremely lucrative proposals that were presented to them very often, and could have enormously eased their lives and contributed to the research.

A Lonely Path

In our materialistic age, it's not easy to talk openly about Instrumental TransCommunication. Everyone broaching this sore subject gets nothing but scorn and derision. (Mockery and derision are the downfall of the world.) In nearly all social circles you frequent, you'll get reactions different from the people's usual behavior when you begin discussing a particular topic:

> Generally," says Hans Otto König, "you are seen as someone strange. People often don't want to hear about the topic, which bothers or frightens them. Or they may appear to be interested at first, but then they veer off to another subject immediately, unless they can quickly assert their disdain for spiritual séances. Most of the time, we face incomprehension

and even total rejection of the phenomena. Those rejecting the ideas completely, without having listened to a single acoustic document, are the most numerous.

They travel into a kind of remote reflection and return before passing through the live experience. Others immediately try to minimize the topic, arguing that it hasn't been scientifically proven, so why talk about it? Sometimes they don't want an area they feel pertains only to religion to enter the realm of science. "Boundaries should be maintained!" Often you are considered part of the "Esoteric Clan," having found the lucrative opportunity to profit from the relief of the pain and suffering of grieving people. Incredible as it sounds, it was like I had to apologize to those debating my discoveries.

Why should I be ashamed to speak freely? Communication with other forms of existence is really not an easy subject to talk about. We must be very careful how we talk to our counterparts so we don't scare them. We even risk being thought of as crazy or becoming an "esoteric smoky" to rational-thinking and righteous people. As long as one stays in the realm of intuition, belief, mediumship, and so on, practicing this activity isn't dangerous. There may be a smile, but no hostility.

But when you enter the area of research called "scientific research," in that arena research yields new information and learning based on experimentation with real facts—using acoustic materials, realistically measurable and identifiable. This new information is reproducible, even if it comes in a different order from what is normally found in official science. Then everything changes, often causing extremely violent reactions, if not dead silence.

Meeting the Media

"Listen: Sensation and Superstition do exist in human beings. They don't know reality."

This message received by the Infrared System becomes evident when we examine the media, which thrives on sensationalism. The media, which disseminates most of the information about ITC, does so mostly fallaciously or disparagingly.

The researcher argues as follows:

Television programs excel in stultifying the subject and building their running commentary to arouse irrevocable doubt about the survival of consciousness. For example, the journalists do this by cutting or distorting important parts of the discourse, reducing the research documentation to a small dialog with a grandmother, or broadcasting only unclear messages while omitting the most audible part. They also interview people whose knowledge is still too limited to respond accurately to technical and complex subjects. They invite those people to perform the experiments, but under such adverse conditions that failure is assured. For example, they can't get back in touch with a person who manifested himself 25 years ago with another experimenter. Why should this entity come back again to meet this new person on the television show? The chance that the entity will manifest again is minimal or nonexistent. It's quite amazing, in fact, how little knowledge we still have about parallel worlds and their interaction with our planet.

Some good journalists do exist in the media, informing in a neutral and objective way. Unfortunately, however, they are very rare accidents, and well-known to all famous experimenters. Friedrich Jürgenson had already warned the explorer, and even after his physical death, Jürgenson wanted to do it again, stating the following message:

> **"Hello here Friedel.**
> **Hans König, hear me; it's me Jürgenson.**
> **Do not forget, on the moon there are rosy cows.**
> **Carry your way to its conclusion.**
> **The crystals weave the universe."**

These words, passed through the doors of death in 2004, refer to a conversation between Friedrich Jürgenson and H. O. König, on one of their many mishaps with journalists. In an interview about their discoveries in the afterlife, they asked this great lover of painting, if even animals lived on the other life planes. The fine man simply replied, "I can imagine it." "Jürgenson believes that there are rosy cows on the moon." This was exactly the title the journalist then chose for his article in the newspaper. That says it all: Print the explanation by consciously reducing and stultifying it.

Meeting the World of Science

H. O. König's foundational research didn't particularly affect the official and approved scientific environment. Nonetheless, in the '70s the study of this phenomenon began booming.

Many people began to experiment in this area, and many more were receiving messages from somewhere, from loved ones and other Spirit beings. So in the '80s, many ITC associations were born in Germany, France, Italy, Spain, and other countries that distinguished themselves in the study of communication with the dead using a tape-recorder or a simple radio.

To believe that scientists would take the phenomenon seriously, on the evidence of a multitude of actual contacts, was unfortunately wrong, as we had already learned in the past. The researcher explains: "As science accomplished both magnificent and terrible things, the history of science teaches us that many discoveries were made long before the official scientific community legally recognized them. Just think of Edison, who was the laughing stock of the scientists at the Universal Exhibition, or Johann Philipp Reis, telephone pioneer in the 19th century, accused of fraud for his invention in the newspaper "The Globe," to name only a few."

Is it surprising that most scientists turned away from the ITC, some even vehemently opposed to it? At the beginning of his research, H. O. König often tried to interest other physicists in the observed phenomena, but he gradually lost his illusions when he met with a wall of silence or a smile of contempt.

Nearly all the physicists he contacted replied that either they couldn't afford to take part in this kind of research (they risked their credibility or being dismissed) or they didn't want to know, and they kept their distance before even considering the phenomenon. Or they considered this research impractical because it couldn't be used for socioeconomic purposes. Even those who were initially interested departed once they realized the disastrous financial impact such research would have on their lives and their families.

In 1983, after the television show on RTL (a famous German TV station), H. O. König and Rainer Holbe, journalist and TV presenter, contacted one of the renowned Max Planck Institutes about König's ultimate invention, the Multi-Oscillation-System, and its results. They wanted a collaboration between the Institute, the researcher, and RTL. The first meeting went very well, and the Institute's physicist was greatly impressed with the research, the device, and the results obtained. He promised to submit the request to his whole team. The two men couldn't believe their eyes when, shortly after, Rainer Holbe received a letter from the physicist saying that, much to the regret of the Institute, they couldn't take an active part in the project: "What would happen to their reputation if communication with the dead turned out to be true?"

The same thing happened at the University of Münster, Germany, where H. O. König was allowed to carry out certain experiments in a laboratory for a time. The researcher consulted several colleagues in their laboratories: none would officially admit his conviction that consciousness survived physical death for fear of the social consequences. H. O. König can't forget the words of one of these colleagues: "You are like someone who has found a needle in a haystack!" But what's it good for? Over the years, the same scenario was repeated again and again at other institutions. Finally, the researcher decided to relinquish his plans and hopes that one day his results would be recognized and studied by the scientific community. He abandoned the idea of seeing the results of this research recognized and studied by the scientific community.

In a 2007 copy of the magazine *"Die Parastimme"* (*The Paravoice*) we can read the following article by H. O. König that explains the fundamental position of the scientific world regarding the research of the ITC:

> Whoever hears about Instrumental TransCommunication for the first time inevitably asks: 'Why doesn't science acknowledge these phenomena? Why doesn't it integrate acoustic documentation into its special field to distinguish between fraud and fact?' There are probably several reasons scientists and the educated public ignore the field of ITC. But all of these points are more symptoms than real causes. These must be deeper. But where can they be found?

> Unquestionably, there is an instinct that forces us to reject the unknown and the inexplicable, regardless of the probable evidence that supports them. Our concept of the probable is normally based on what is usual and known. It is determined more by feeling than by reason. Imagine someone tells you that he has met a dead man with whom he had a pleasant little chat. I'll bet most people would laugh at this statement as absurd. But this attitude entirely lacks understanding: No one can theoretically know if such apparitions exist or not.

> We could ridicule this story or simply add it to the list of miracles for the simple reason that it is far from what is usual and known in our European World. It

is a familiar part of human behavior to assess likelihood in the light of what is familiar. It is well known that many people laugh at the behavior of foreigners, simply because it is unfamiliar to them. One day a Frenchman told a Hawaiian about children running on a frozen lake in winter. The man doubled over with laughter: the sight of a group of little boys and girls, running on the ice, was really too absurd for him!

In some way the interior, familiar thought pattern of the Hawaiian was confused. But we should be careful not to make fun of him, because brilliant minds, highly intelligent and cultured persons, renowned philosophers, and people of all kinds behave exactly this way when talking about ITC. Here the educated man, just like the man in the street, is likely to be seized by emotional impulse. Here is proof, evident in some quotes received from scientists in response to my writings.

It is useful to know that none of them had ever heard about an acoustic document and no one bothered to investigate the phenomenon. Here are their comments: "Your presentations and your research results are based on fantasy coupled with a lack of critical thinking and philosophical knowledge"; or even, "If Instrumental TransCommunication really existed we would apply an irrational world at the expense of a rational world"; and finally, "Mr. König, your so-called spiritual communication with the dead is as low as the voice of spiritual entities you claim to hear."

These intelligent gentlemen, whom we would believe normally act with a scientific mind, actually expressed prejudice. Prejudice among scientists seems to be the main characteristic of their general attitude. In my opinion, certain facts have no right to exist because they confuse the rational order of things. So why is an avid enthusiast of the exact sciences interested in the fundamental research of ITC? I will now quote a remarkable scientist: "Even if ITC does exist, we have to repress and hide it. It would destroy those premises without which real science couldn't continue its work . . ."

"Of course," the researcher goes on, "there have been many exceptions. Here we should mention many names, scientists like Professor B. Heim, Professor A. Resch, Professor Bender, Professor Schiebeler, Professor W. von Lucadou, Professor R. Chauvin, Professor Senkowski, and more."

O. König recalls, "Along the way I had some very interesting conversations with many mentors. Those with Professor B. Heim were thrilling. His hypothesis of the twelve dimensions opened a new way of interpreting the world that challenged me. Then, of course, there were my many exchanges of views and ideas with Professor Senkowski, an intellect of extraordinary eloquence and a man of great kindness. He contacted me after having listened to one of my live experiences in a radio broadcast."

Unfortunately, he never started fundamental research of the phenomenon, nor did he develop new systems, but he had great experiences and a broad knowledge in ITC because he experimented himself and gave himself the mission to follow the research of the experimenters closely and make them known, both in Germany and elsewhere. At one point, however, our paths of reflection and experience separated for various reasons.

Now let's look on this side of the Atlantic. We see that the United States was much more open to this phenomenon, in full swing in the early 1980s. Some institutes for scientific research personally contacted H. O. König. In 1984, the University of Atlanta even gave him the title of Dr. h.c. for his research in TransCommunication. Several people wanted to collaborate with him, but as he says, "Generally speaking, interest was mainly subjective and the proposals not particularly interesting: my results for money!" Even so, in forty years the world has changed, and science has evolved.

From 1984 we move on to 2014. It seems that in the last 15 years many ITC experimenters came into the spotlight, like Anabela Cardoso in Spain, Marcello Bacci in Italy, and Sonia Rinaldi and Carlos Nunes in Brazil, surrounded by teams of scientists. The researcher followed with interest the scientific developments and the quantum physicists who arrived at amazing research conclusions:

> They tell us that everything that, at a given time, was connected always will be, and that matter is not as dense as we think. It is all energy, just as consciousness is all energy. In fact, the man who looks through the microscope is as important as the microscope itself.

The researcher looks forward to further discoveries. He is convinced that one day science, having already made so many discoveries about the power of thought, will also recognize that consciousness survives physical death:

> It's only a matter of time, because matter is nothing other than densified spirit. Our world is a materialization of thought patterns. Just as there are waves of heat, light, and electricity, there are also existing waves created by mind. Thoughts have a tremendous power that we could use so much better if we had a greater knowledge of the workings of the mind. And the powers of mind will be intensified by the increased understanding and recognition of Mind.

After a pause, he adds,

> There is a whole series of filters between us and reality. The first filter is one of our five senses. For example, a blind man cannot see the color red. That is a biological filter, but there are also many psychological and mental filters. So, if we have a mental representation of something, we can see what is going on in that representation. We do have a certain inner expectation or attitude about something, and therefore will experience exactly what corresponds to this expectation. There are thousands of things we would

be able to perceive, yet we can capture only a small part of these, those that fit our paradigm. We get to see and to recognize only what we already know.

As for the questions the explorer asked himself in 1974 (What is the nature of this phenomenon? How and why does it exist?) he is still questioning today. Even though he has received certain answers during his forty years of research, he is well aware that we are still at the beginning of this research, which opens up to us endless possibilities about supernatural worlds. He goes on to say:

> At a minimum, I should live for another three hundred years to achieve all the research projects I have in mind. But my body, in which I'm really starting to feel cramped, cannot cope with this challenge for as long as I wish. So I absolutely must reflect on what might still be feasible for me, before stepping down one day.

A Way to Live Death Otherwise

Does the explorer fear death? He smiles:

> How could I be afraid of something that has become so familiar to me? Furthermore, my curiosity is stronger than the fear to leave what is known to me. I want to learn so many new things that are impossible to witness as long as I am connected to a physical body. At first-hand I will experience the release of my physical body and perceive it as true salvation. Thanks to my ability to enter a trance, I know a thing or two about that feeling at such a moment. So I am no stranger to being in the realm of Spirit.
>
> I know that I can walk away when the time is right, which is nothing special. Many people, like yogis, know how to leave this world at the decisive moment simply using the power of their mind. I have

the great opportunity to wander between the two worlds at length. And then, one day, my spirit will stay there and not be reincarnated anymore. That's all! Certainly there will be some transformation, but not my obliteration. I do not think that dying will be difficult for me, but it will be difficult to leave those who are dear to me and for whom I am still responsible here below. Leaving them for good will be painful, but not for me personally.

Perhaps if one day we know more about what death really means, the pain of loss will lose its violence and death, its terrifying nature, which is attributed to it so often in our world.

Through their volunteer activities and research, H. O. König and his wife observed how TransCommunication can help a person experience death with more serenity and give precious help to people in grief during the loss of a loved one. Over many years the couple accompanied many grieving people or diseased persons through their dying process. Very often they were present at that very moment when someone, young or old, was passing through. In this way, they came face to face with death, sickness, and all the accompanying feelings, like pain, fear, anger, rejection, and despair. It is not easy to cope with all these realities, and maybe we will never get used to it. "Since I've been working on the ITC," König recalls, "the phone rang so often, and I could feel the tears before responding. The beings could no longer say a word."

Such as that mother, who suddenly lost her whole family in a freak accident. She, her husband, and their three children, three, six, and eight years old, were taking a car trip. The morning was very foggy, with very poor visibility. They wanted to stop a moment. As they were hardly able to see anything, the woman got out of the car to indicate the side of the road to her husband. As soon as she was out of the car, a truck crashed into it, and it caught fire immediately. Before the young woman's very eyes, her husband and her three children were burning to death, screaming for help as the car went up in flames, while she stood powerless to do anything for them. The pain in such a moment is so enormous, and the question "Why?" incessant.

"How could that happen? Maybe the driver was asleep at the wheel? Why was I the only one to get out of the car? Why not the children? We were so happy after the birth of our three children. Every morning I wake up with the same images in my mind. He, up in heaven, why couldn't he let me die with my family? What is the point of all this?" The screaming of the soul is tremendous. So the best thing we can do is listen, be there amid that pain. "And you are still talking to the dead!" That exclamation is never thrown in my teeth so hurtfully as in such moments. I always keep silent, trying to lend an ear, often for hours. Then I say softly, "I don't talk to the dead, I'm talking to people who have left their physical bodies and are now in other dimensions. I'm communicating with them from mind to mind, with their spirit, which is in another reality . . ."

The researcher is recalling the past, remembering so many names, faces and destinies, their questions, their doubts, and also their great transitions. Such as that mother, sick and speechless with grief since she had lost her three children, three boys, all within a month. She had been interned in the hospital, where this time the psychiatrist himself contacted the researcher. The doctor didn't believe in the afterlife, but he was ready to make the same trials he had heard about on television. Accompanied by his patient, he visited the researcher, who agreed to try to do a contact for this woman, eaten up with grief. The contact was established.

One of the sons announced his first name and said, **"Mamma, we're all fine,"** and **"Happy here."** Suddenly the fortress of pain the mother was walled inside began to open, and gradually she found her way back to life. We can carry experiences like this for a very long time before they make their way into our inner Self, finding a shelter in peace. During all those years, the researcher and his wife found that the personal experience of tangible contacts with spiritual presences from Another World contains great transformative power for many people.

It's an indescribable experience hearing, for the first time, a name or a simple **"Hello."** On the face of it, the content is completely ordinary, but the reality of establishing contact with the spirit of the deceased person is irreplaceable. It allows you to feel, in your own being, the evidence that death does not have the last word. I call someone who responds to me, standing on the other side of the river of life. He tells me his surname, or his first name, or he calls me by my name. He wants me to know that he is still alive, that he is still connected with me, but even so:

"We call on you, everything is OK.

Believe, Diana is alive.

Yes, you hear me. I see you."

These palpable words often carry the person to other worlds of thought. There, where violence does not exist anymore, but passage, transformation, and continuity, the espousal of death with life, two faces of the same reality that is life. The soul of the one who is listening is deeply transformed, touched by the wings of love, skimming for an instant the body of the one who, some day, will join this other world with its invisible inhabitants. Even if, for the one who remains, this subtle touch is certainly not comparable with the communion of two physical bodies, resting in these unknown spheres he has the power to arouse in himself a new knowledge.

Even if our need for consolation is not satisfied, new seeds, and especially a strong inner peace, bloom through this. When the perspective on life changes, other interests germinate and flourish inside. The death of someone, as paradoxical as it may seem, often means birth for another into a different form of life, lived as a possibility of transformation and inner growth. At the same time, the depth of the wound stimulates—with the pain of childbirth—a rebirth of oneself, or growth, often against one's will at the start, then sometimes with acceptance, and consciously.

When we are supporting people with a terminal disease, the discoveries made by ITC can be very precious. The researcher remembers vividly a very young man he often visited in the hospital at the request of the director. Peter refused categorically the idea of dying and didn't want to know anything about the priests who came to visit to help him. He didn't accept his terminally ill situation, refusing to believe that he would end up in the loving arms of God, as they taught him. The researcher had very long conversations with this young man and made him listen to many recordings of the electronic voices.

Little by little Peter began to accept his destiny, to prepare himself for the great transition. Right before he died, he repeatedly asked the researcher to find a song for him he particularly cherished. It arrived just in time for him to listen to it until his last breath. Peter died with a smile on his lips. The title of the song was: "Never We'll Die at All" After the funeral, the deceased young man got in touch with H. O. König, saying: "**Here is Peter. All is as mentioned earlier,**" referring to the song title.

For all those who are faced with the inevitable, this knowledge, discovered in such a tangible way, lets us vacate the world's stage more calmly, without inner battles, but with the certainty that the adventure of consciousness goes on, with new experiences of life. They know there will be someone who awaits their arrival on the other side of the screen, opening new areas in and around them, filled with love and belief.

A Path of Interiority and Knowledge

The researcher, however, always insists that the purpose of Instrumental TransCommunication is not to establish daily contact with our loved ones. Quite the contrary, it is essential to leave the others completely free and not to burden them with our personal grief. It may even become a problem of the mind to whom we are connected: **"Your sorrow is our problem."** This message appears often. Many people then retort, "Why do you spend so much time contacting and getting so many painful radio messages if not to contact your relatives? Why waste so much energy and time to receive only some murmuring or a few words? Would it not be better to focus on other facts of life?" The investigator replies,

This question surprised me greatly, because I am fascinated by what is still inexplicable and of anything that might give a vaster knowledge about our universe. Apparently, these people are content with the knowledge they possess, their beliefs, and the life they lead. But ITC could prove itself so valuable, not only for the researcher but for everyone.

As the knowledge transmitted to us is often inadequate and too restricted, man feels the need to move on to find more satisfactory answers. Sometimes, to do this, he travels the globe, studying books or seeking the answers in himself. Science doesn't tell us much about death, so those who want to know more have to look elsewhere, whether in religion, esoteric schools, great spiritual traditions, the practice of Buddhism, or other methods. Occasionally he also encounters ITC.

Thanks to direct contact, he can hear the link that exists beyond the visible. This acoustic footprint of the link can be a key experience in realizing the permanence of personal consciousness and encouraging others to walk on the path of this knowledge. For those ongoing recordings mark a personal road opening onto a space where we can try to establish a connection in us between consciousness and Superconsciousness, which is always connected with the Invisible Worlds.

This path into our Inner Self connects everyone with an immense power, and the different steps we pass through are very rewarding. One day you may encounter that entity that has assumed the duty of accompanying you in your life, your Guardian Angel or your Spirit Guide, who has known you since birth. You will feel his love guiding you by various inspirations when the time comes. Personal meditation will become a privileged moment in your daily life. Conversations by thought begin to develop between you and the entity personally affiliated with you.

You can feel profoundly that they accompany you in every situation in your life. Little by little your sensitivity sharpens, and new abilities are awakened and new interests are born. Your review of humans, animals, and plants changes dramatically. The joy of discovering your Inner Self, unfortunately often unexplored, is intense, and the fruit it bears will give you bliss.

"I realized," the researcher says, "that every communication strengthens my own relationship to the Invisible World, a personal and highly confidential communion, growing and thriving all by itself. Small whispered words are often the most touching ones, bearing witness to this relationship from heart to heart: "**Do you hear me? I come tomorrow at 7.**"

And the next day, at 7 p.m.: "**I am here: Do you hear me?**"

"From these moments of real sharing, that relationship can begin to intensify. It seems that in our Inner Self, a huge unfathomed area opens and grows with each contact made beyond the boundaries of our senses. These contacts are like seeds that are sown. Doubts, concealed in every one of us as to the survival of mind after death and the existence of Invisible Worlds, gradually disappear due to the reality of these tangible experiences. Slowly we become one with the Spirit World. The belief in an afterlife turns into knowledge, and Our Self flourishes with regular contact and the intensifying relationship with our new invisible friends. Thus far in the experiments, we find that each transmitted word can be significant. All these little words help me see life differently, to identify the Parallel Worlds and build all my great devices for contacting them."

The path of making recordings by radio is an individual path and a unique experience of direct contact. All the words received are pearls that testify objectively to your own link. They are like white shells containing small diamonds that reflect the world of the beyond. We are forced to leave the noisy world that surrounds us to relearn how to collect ourselves, to concentrate, and to listen to what is so fine, so silent in this same world and in ourselves. We move away from the outside world to pray and to open up a path—a unique personal voice in communion with another voice, coming from the Invisible World.

The researcher explains: "I think that we do not have the right idea about the value and the meaning of these radio contacts, expecting that they will give us extensive information about the Afterworld. This is, incidentally, impossible, given that the necessary technique is lacking. But if it doesn't provide information, it allows everyone to establish contact all by himself and even intensify it. It is an opening for countless entities to get in touch, thanks to its very wide frequency spectrum."

Meeting Other Paranormal Phenomena

During his pioneering, the researcher became aware of many paranormal phenomena, both through personal experience and through stories he had been told. Some are so amazing that he never wanted to talk about them, for fear of endangering all his discoveries and his basic research work. He explains:

> The ITC work is already so revolutionary in our materialistic age, requiring a completely new mode of thinking and an open mind. It's like diving into the Unknown, which evokes many fears and is therefore rejected. But today, at least, everyone can experience this for himself—listen to the sound documents that exist and attend the spectral analyses. As for all the other incredible stories, please take my word for them. It's just a question of belief. Most people don't have personal experience of all these different phenomena. Because they're unable to make the necessary distinctions, they are confused and find nothing at all.

In fact, they prefer to remain silent rather than risk discrediting the progress of ITC. The researcher remembers the many times he was called to intervene in haunted houses or in situations where chairs moved, devices shifted, or objects fell without any human intervention. He observed unmagnetized scissors spinning around in his lab when he asked if someone was in the room. Or tape recorders turned on by themselves, without being touched by anyone.

He witnessed an old telephone, wireless and unconnected, ring without reason. He himself experienced some materializations, saw many apparitions, and received paranormal pictures with incredible clarity. Each time, the researcher tried to find explanations: first the natural and then the supernatural ones. According to him, in many paranormal phenomena, animistic theory is in place. In every being, there are hidden inexhaustible forces that can be triggered in certain circumstances, causing unprecedented phenomena.

Thus, the story of a family, which he had visited because they experienced some very disconcerting events. Books regularly flew across the room, or objects shifted by themselves. The researcher asked the family what, when, how, and in whose presence all this happened. Then he realized that these phenomena manifested only when the daughter, a teenager in puberty, was in the room. Unconsciously this girl, feeling great inner tension, was the origin of all these phenomena. Once he had a long talk with her, explaining some things, all the strange phenomena stopped immediately.

But in other cases, the animistic theory proved insufficient, as in the following example. One day König was called by a young woman, the tenant of an apartment where really disturbing things were happening. The woman was about to move because the situation had become so unbearable for her. The researcher went to her home, but couldn't discover the reason for all the strange noises.

He returned to the apartment with his wife, and together they began trying to make contact with the device. During the first attempt, a voice came through, shouting: **"Get out of my flat!"** The researcher visited the owner of the house, hoping to get more information about the place. He discovered that the previous tenant had terminated his lease quickly; and before him, an old man, who had lived there all his life, died of a heart attack while watching TV. The couple attempted another contact, inviting the spirit to go toward the light and follow those who were waiting for him.

The deceased person replied, **"But they are all just dead."** The couple tried to converse with the spirit, who still hadn't realized that he was dead and thought the young woman was trespassing in his home. Unwittingly, the man experienced such a force that it was causing terrible blows, fearful noises, and moving chairs. After that intervention, the spirit finally moved to a plane of another life, and the woman got on with her normal life.

Interest in the Human Psyche, Hypnosis, and Trance

This German pioneer is passionate about everything related to human beings. For nearly thirty years, he was an enthusiastic subscriber to scholarly journals, and even today he closely follows progress in neurology. But that didn't suffice—he wanted a solid understanding of the human mind by other means. So he and his wife studied for three years at the University of Zürich. For many years after that, they ran an important private practice in "hypnoanalysis." According to the researcher, "Hypnosis is a marvelous instrument for reaching the subjacent stratums of the psyche and finding out the root causes of a problematic situation."

During various hypnosis sessions, the everyday consciousness shrinks to the point of reaching the subjacent spheres of consciousness. Stepping back into the patient's life often reveals the roots of a problem and even the causes of diseases. The sessions finish with analysis, followed by therapeutic conversations, ensuring a healing process in the patient, mostly with very good results. Thanks to his daily practice, the researcher could connect the dots of the human mind, gathering data on different types of people.

The researcher argues, "The human being is so complex that no psychological education will ever unravel its mystery." He also wanted to examine more closely the following question: "What influence does the psyche of the experimenter have when contacting the World Unseen? Can each experimenter make contact with any entity in various parallel planes? What type of person receives what type of answers from the Spirit World? Or does the human mind itself cause all these phenomena?"

To better understand, he went to a psychiatric hospital in the region. Every two weeks for two years, he visited some patients who claimed to hear voices and see spirits. Was their reality the same as the one the researcher knew? What was the origin of their current reality, which was considered mad? He observed, made notes, and analyzed different cases, all while acquiring new knowledge. Then in 1984, he founded the FGT *(Research Foundation of the TransCommunication)*. This community, and all the events and adventures he had with people involved in some way in the mystery of the afterlife, formed his great breeding ground for practicing ITC.

In addition, by practicing hypnosis with the help of his wife, he personally experienced deep hypnotic states where his mind was completely free of his physical body. He learned to recognize the duality of body and spirit despite their connection. In these subjacent states, his soul entered a subtle world where totally unknown information was revealed to him. He became aware of his Inner Self, "this thin film of the human psyche" existing beyond time and space. These altered states of consciousness are called "trance" by many indigenous peoples. Man is actually capable of walking between parallel universes that intermingle.

In fact, the trance is a natural human state, although little is known about it. In this altered state of consciousness, the researcher received many messages and could observe materializations of phenomena, such as a silver ring with a blue stone. One day he held a séance with a small group of interested people. When he was in a state of trance, they began questioning him about the power of the mind. Suddenly, a ring materialized in the room and settled on the chest of the researcher. Offering the ring to his wife, Margaret, she vigorously refused it, remarking, "It is not for me. YOU keep it!"

After trance sessions, the researcher often said, "I wish I could use a camera to photograph all these fabulous images that I discover by browsing these subtle planes of the spirit, freed from my physical body, though still feeling a certain connection with it by a thin rope. My time to stay in those worlds, however, seems to be limited. The two entities who are accompanying me on my journeys always tell me when it's time to leave and join my physical body again."

The Origination of Supernatural Faculties

Very often, the researcher retires to meditate. He deeply regrets that meditation is still outside the scholastic curriculum. "Our society," as he explains, "is extremely chaotic and, for the moment, too fast. How do human beings still have time to live their own lives? Most people are exhausted and no longer live themselves. Often they don't know who they are or where they are going. Direction is missing in their lives. For "Taking-the-inward-turn" you must take time. A deep connection cannot be established in a few minutes. So today's consumer society and its social surrounding is subjected to undue haste and a mad rush. What's the hurry? What do we ultimately gain?"

The researcher likes to take it slow. This is a man who goes against the tide, living his life different than our society dictates. But doesn't going against the tide take us back to the source? At least, we conclude, we get a better idea of a life based on "being" and not "having." Some define him as a wise shaman or a medicine man of modern society. Others take him to be their spiritual father or an Elder of a Native tribe.

He, however, wants nothing to do with these names. Yet many contacts with the Unseen have awakened in him psychic abilities, which in most people can be evolved only during their deep state of sleep. He often sees the etheric bodies of the deceased materializing in our world. These apparitions, however, always happen spontaneously: phenomena that cannot be prompted by anyone or anything. He remembers, for example, an afternoon in Wolperath, a city in the northern part of Germany, where he was giving a lecture on ITC. He saw an unfamiliar married couple entering the hall, accompanied by their dog. He welcomed them, remarking, "What a cutie, your dachshund!" The woman was stupefied. H. O. König saw their dead dog with them, and even knew his name. His description of their dachshund was exact.

"If I told you everything I know," he says, "you'd really think me mad as a hatter or an inveterate dreamer!" All these phenomena seem incomprehensible, even absurd, for a rational thinking mind. Isn't that rich? "How is it possible that the etheric body materializes within a very short time, becoming quite indistinguishable from a physical body?" Once again, he can only construct a new hypothesis, being unable to verify it.

A neat remark: "Most of the time, astral bodies don't manifest clearly and distinctly."

This man, connected continuously with the Spirit World, developed another ability: directly capturing the thoughts of the person standing in front of him. To do this, it is essential for him to connect with the other person, to enter into a communion and no longer see his appearance. Then, suddenly, like a thunderbolt, a number of allegorized thoughts come from the other person into his own mind, as images. Later, he developed many other abilities of psychokinesis or telepathy, but he considers them all insignificant.

The desire for knowledge is what characterizes this man most of all; he continued on his independent life path, experiencing new things every day, anxious to deepen his research. He is always willing to learn more about life, existence, the human spirit, and the possibilities of communication with the Unseen. His great passion is spreading knowledge based on direct experience with the Spirit

World and then sharing it with others. To do this, he is now accompanied by that life partner who wears the silver ring with the blue stone that he received during a séance. His grandfather had even mentioned her when he was 20 years old:

"Hans Otto, someday you will meet a woman. No, I'm not talking about Margaret. She will play a very important role in your life." Then his grandfather revealed her name. The researcher remembered this when, years later, he met a young woman, well after the death of his wife, Margaret. The woman in spirit, now living on a new life plane, came through during one of the researcher's trance sessions. She welcomed them both with the following words: **"Communicate to the humans. I was aware, I am aware, I'm still here. I'm still me; nothing has changed on this level."** And the researcher added, to the best of his belief:

> Our consciousness survives physical death. Too many clues indicate this. I'm convinced that the day will come when human beings will understand that consciousness is much more vast and connected, with more worlds than we can imagine. At that point, we will understand how primitive and mechanical our traditional science is, how ridiculous it was to believe that consciousness is only a product of the brain that dies with us, even though a precise definition of consciousness is still lacking. I don't communicate with the dead or the deceased, as is often mistakenly believed. I communicate with other forms of existence, with more subtle information fields, existing in parallel universes.

Second Part

Itinerary of a Singular Research

"All truth passes through three stages. First, it is ridiculed. Second, it is violently opposed. Third, it is accepted as being self-evident."

~ Arthur Schopenhauer

Starting to Discover Acoustic Phenomena

A television program in 1974 was Hans Otto König's trigger for starting overwhelming research that would radically alter his own worldview and force him to go far beyond the boundaries of his physical knowledge sanctioned by science. It all began one autumn evening when he was watching a controversy on German television over an acoustic phenomenon titled "Paranormal Voices."

A man named Friedrich Jürgenson claimed in front of millions of viewers that he had received, on tape, voices of deceased persons who spoke to him and wanted to be recognized. He had captured these voices accidentally while recording bird songs. Hans Otto König, an electroacoustic enthusiast, heard the murmurs presented as coming from the dead, but he did not understand their origin or the manner in which they had manifested on tape.

A realist and physicist by training, he was convinced that these fragments of sentences came from the unconscious of this distinguished man and certainly not from his deceased mother, as that man on television dared to assert. But what an amazing discovery! This young German had never heard of this phenomenon. So would there be a way to capture our unconscious thoughts? Would man be able to record and hear directly the silent world of human thoughts? At all costs he wanted to learn more about it and examine the mysterious phenomenon of these voices that were called "paranormal," signs of another world, invisible and reputedly silent. What was really the nature of this extraordinary acoustic phenomenon? How did it work and why was it there?

He asked himself question after question.

The First Voices, Received by Conventional Methods

To acquire a well-founded knowledge about a topic and to form a judgment about it, it is necessary to study it thoroughly and then experience it yourself. To learn more, H. O. König extensively acquainted himself with the existing recording methods—the simple radio method, with background noise like trickling water, or the procedure using a simple microphone—and furnished himself with the necessary technical equipment to run multiple trials.

He went on to meet other experimenters, intensively studying their methods of work, their thoughts, and their results. On several occasions, he had the pleasure of meeting the man who had instigated his entire research adventure, Friedrich Jürgenson, to discuss this paranormal phenomenon where the Spirit World interferes with ours to speak to us through technical devices. They discussed one issue, for example, at length: What is more important when making contact by radio, the technology or the experimenter?

And here is where they parted company, after wonderful times of reflection and shared experiences. Friedrich Jürgenson was convinced that the radio played the main role in the recordings, whereas H. O. König thought the experimenter took the lead. Unfortunately, H. O. König didn't have the chance to meet the second renowned pioneer, Konstantin Raudive, but he became acquainted with his work. So he was able to get an idea of the average sound quality of an electronic voice, which is, sad to say, most of the time extremely low. He even discovered that in 1974 the Latvian-born experimenter was the first to submit himself to a very rigorous study, in England, under the most stringent laboratory conditions, which excluded the theory of interference with other radio frequencies and appeared to demonstrate clearly the paranormal character of the voices. The researcher, however, was still very skeptical. He wanted to study in depth these voices whose origin he could not understand.

So his first experimental laboratory was established in Ratingen, near Düsseldorf, Germany. In this town of his childhood, Hans König tried all the conventional recording methods. But his sincere desire to understand this still unknown phenomenon was put to the acid test.

He ran a test every Tuesday night for a year and a half before managing to hear a single electronic voice. After persevering so long without success, he made the following comment one evening: "I've been trying for so long and have heard nothing."

"It's on it. The old tape."

Finally, he clearly heard a voice! A correlative response to his question. The old tape? Which old tape? Twenty tapes had already been filled with recordings. The old tape? Wouldn't it be the one that was different from all the others, the one he had used during his very first trials a year and a half ago? He pulled it out and for hours, inch by inch, listened to the black magnetic tape.

In doing so, he forced himself to learn by the slow process of listening to and deciphering electronic voices, which is often difficult. Little by little he realized what to listen for. He remarked, "Learning the language of paranormal voices is like learning a new and unknown language." During these many hours of intense concentration, he slowly learned to distinguish, from all the radio noises and other voices, fine whisperings. He distinguished several words, like **"I am alive, help me, pray for me."**

And, finally, he heard the following message:

"Uncle Hans comes to visit."

They were whispered words, like an expired breath. You can imagine the joy of the researcher. He had finally been able to hear a significant sign: a name was transmitted and a message with an important statement about his questions. So during his nightly recording, H.O.König asked: "I've heard you, Uncle Hans. Do you know, who I am?"

"Hans Otto."

The "Uncle Hans" referred to had fallen ill shortly after his nephew had made the fortuitous discovery of the paranormal voices. At the hospital, the young man had told his uncle about his discovery and his desire for research. The uncle had simply replied, "If our consciousness survives physical death and the opportunity of being heard exists, I'll be the first to give you a shout. But honestly, Hans Otto, I agree with you, I'm not convinced either!"

Where did that message come from? Was it the consciousness of the deceased uncle that had spoken to him? Would he fulfill his promise? Was it a sign of his presence, still alive despite his passing? The researcher doubted it strongly. On the contrary, this message seemed to support his first idea: the voices are the expression of his own deep desire, and therefore of his unconscious thought. But how was his unconscious thought able to affect the soundtrack?

He proceeded with his many recording trials in the ensuing months and years. He continued to observe and evaluate the data received to better understand this phenomenon, so new, unique, and fascinating.

What is the nature of the phenomenon?

He intensified the contacts to up to three times a week. What parameters interfered with the existence of these electronic voices and with the contacting? In what area of technology did the voices manifest themselves? How did they get on the tape? What was their origin? What were their characteristics?

Day by day he tried to get a better idea of this phenomenon, which contrasted so much with his physical knowledge. He did this not by constructing working hypotheses he couldn't verify, but by experiencing and systematically studying the different elements at play: the technique, the personality of the experimenter, and the received messages. Each element of the technique was studied separately in his laboratory. To detect what was happening during the contacts, he examined everything from the microphone membrane, heads of the tape recorder, and conduits to the magnetization of the tapes.

He experimented for a long time with different radio waves: short waves, medium waves, and long waves. When and where did the contacts occur and prove to be of the highest quality? What differences showed up? Then he acquired various kinds of professional microphones, highly specialized and extremely expensive (infrasonic, for example), with high-value capacitors and later, ultrasound. Were there significant notable differences?

He kept specific protocols for these experiments to detect, among other things, how much his own mind influenced the contacts. At each tape recording, he inserted a description of his emotional and psychical state. Sometimes he made contact while trying to solve difficult mathematical equations or while consciously putting himself in stressful situations, always for the purpose of observing and understanding the phenomenon. Every day he made new discoveries. It was an exciting time for the researcher, who had always adored breaking the mold.

Sometimes he met with a beautiful white stone, like the evening he began a contact after a strenuous day with many problems from which he couldn't break free. During his trial he heard the words: **"I do not understand you."** Then, **"Your thoughts are too confused."** Years later he received a similar response when he asked, "Why is it that some people have very little contact?" **"Their brains are too excited. Agitation is not useful."**

He made many trials using several tape recorders installed in different rooms, running them at the same time on the same frequency. He himself was present, of course, at only one device. He compared the different recordings. Were the voices on all of the tapes? Then he checked the geophysical conditions, received by the bulletins he could access through his amateur-radio license. Did they have an influence? He always noted the lunar phases and the eruptions of the sun. For several years he observed, compared, and analyzed his results. He often had his results verified by several independent persons to see if they could hear and understand the same words he did

Some Characteristics of "Radio Messages"

Following this procedure, slowly but surely he detected some typical characteristics of these voices thanks to a simple technique of mixing different radio frequencies, foreign language waves, or the sound of a radio station to carry the contact.

The contact itself was recorded on a magnetic tape or a cassette, later replayed. Using this technique, he received very short messages, and often only a few scraps of conversation, not exceeding 14 syllables.

The sound quality was highly variable. Some recordings were clear and audible, but most only came through like a faint breath, a whisper or an exhale, like a sigh. Sometimes they were spoken quickly or, conversely, the vowels were drawn out slowly, which required a tape recorder that he could accelerate or slow down while listening. Sometimes the diction was a little robotic. All these properties were partially related to the technique being used.

After having examined them closely, many discoveries allowed him to argue that the communicating entities themselves supplied a frequency that briefly contacted the frequency provided by the radio. These two frequencies created interference. The modulation of thought patterns coming from the communicating entities was manifested on this interference.

The provider during the radio recordings is never stable, so the frequency of the interference is very short. Therefore, the modulation of their thought must be done very quickly with this method. H. O. König often offers the following image to clarify this unfamiliar data of frequencies and interference: "It's like two motorists driving at high speed on a crowded highway, trying to drive parallel in order to join hands through their opened car windows. It can only happen very briefly."

Two examples illustrate this teamwork between the visible and the invisible communicants:

H. O. König: "I'll try another frequency."

"So I'll try it too."

H. O. König: "Which is the best frequency of the three I'm offering?"

"Let's take the high frequency."

During his experiments, however, each investigator must himself discover the best frequency for contact. There is no rule or recipe.

Sometimes the messengers themselves convey the frequency necessary at that moment for the contact. These frequencies are variable. H. O. König one day received the indication for short waves of 10 MHz or 7 MHz. This kind of opening, which is particularly adapted and permeable, the communicating entities call **"Frequency Window"** (Frequenz Fenster).

This example leads us to linguistic characteristics: **Frequency Window** is a typical term of paranormal language. The Spirit World often express their thoughts in images, with numerous and diverse levels of interpretation, which everyone can interpret in their own way according to their intelligence, culture, sensitivity, and experience. Listen to other examples:

H. O. K.: "What is important for making contact?"

"Persevere a ring month."

H. O. K.: When do you get in touch preferably? Can you tell me that?

"By ring-touch."

H. O. K.: "Don't the contacts disturb you?"

"Contact and ever Sunday."

H. O. K.: "What can I improve to make a longer contact with you?"

"The crystals, with meaning in the appearance."

H. O. K.: "What frequency do I have to use for contacting you?"

"You have contact on 937 and the modern technology no longer than 3 hours."

Every person can choose how to interpret this pictorial language according to his own understanding. Each one must decode the puzzle himself.

Other interesting characteristics of their language are archaisms, neologisms, and erroneous syntactic constructions. For example:

"Here comes a new life."

"Norbert speaking."

"I have the proof, contact born out of death."

H. O. K.: "What can I do for you?"

"Help us to the truth."

H. O. K.: "It is difficult to establish contact with you only by the Infrared System. Have you chosen A. M. Wauters for that?"

"Anna-Maria, be welcomed by me."

The invisible communicators have great difficulty transmitting their reality to us. They manipulate the language to create new realities corresponding to their thoughts. As they said: **"We must learn to speak here."** These difficulties are quite understandable, considering that they're communicating by pictorial language, calling them "thoughts."

"Have understanding of the difficulty of the communication with us."

or

"Have problems with voice modulation."

It seems that even they have to learn how to talk with us about technique. **"We must learn here to speak again."**

This difficulty is easily understood if one bears in mind that the invisible world communicates in pictorial images, which they like to call "thoughts." Then, in 1979, the researcher found another white pebble, sparkling with clarity. Thanks to the many spectral analyses of paranormal voices conducted in his laboratory, he observed that the basic frequency was missing in all the voices he had received. The latter is typical of the human voice because it is linked to our vocal cords. From this major discovery, he was able to analyze and objectively show the profoundly different nature of the paranormal voice compared to the human voice. The voices surely had a different origin, but what? And where did they come from?

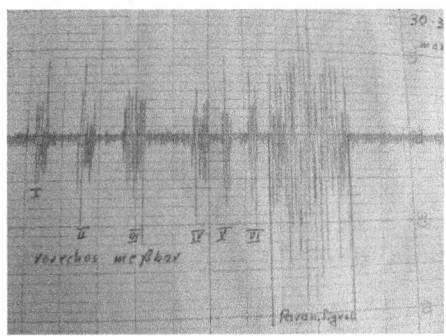

Paranormal Signal

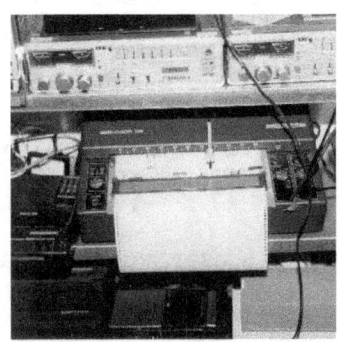

Registration of a Paranormal Signal

Who or What Affects the Audiotape?

The researcher continued to ask himself this decisive question: "Are all these messages the product of my own unconscious, or is there another reality in which something of our Self continues to exist—or is there even something else?"

On the other hand, he had already gathered many messages like these: **"The dead are speaking"** or **"Contact with the Afterlife,"** **"Here the Afterlife's talking."** He received many names and surnames, always in reply to the question, "If an afterlife does exist and I do not disturb you, I would be happy to receive a sign from you, please!" Or he asked, "If I can do something for you, please let me know!"

To his great surprise, he received numerous cries for help and many requests for prayer from communicating entities. But he was unable to verify the identity of, or the information received from, any of these messengers. So who was talking to him? One evening he had a pivotal experience in this regard when he heard the following message: **"Help me. I would go back to my mother."**

This sentence caught the researcher's attention, and the next night he asked, "Yesterday someone asked for help to return to her mother. Who are you?"

"Gerd Siebenreicher."

He developed a little communication with this invisible interlocutor on the following evenings.

H. O. K.: "What happened to you?"

"Am fallen out of the window, at Christmas."

H. O. K.: "Where did you live?"

"Kaiserswerther Straße 23, Ratingen."

Hans Otto König had never had an exchange with the same entity several days in a row, and to date, he had never received a name or an exact address that he could check. Personally, he had never known or heard of Gerd Siebenreicher. The street, however, was in Ratingen, in the same district where the researcher was living, so he went out that night to verify the address. What a disappointment when he found no house at number 23, only a big flower garden. What could it all mean? Where did those messages come from and why did he receive them? Was someone trying to play a bad joke on him?

The next day he went to the town hall to find out if Gerd Siebenreicher had lived at that address. But having no date, he couldn't discover anything. And why did this same Gerd Siebenreicher tell him, that same night,

"Gerd Siebenreicher always come back on your tape recorder."

His wife, Margaret, had the idea to talk to the regulars of an old Ratingen pub. Luckily, an old man with a pipe in his mouth lived in Kaiserswerther Street. He could remember Gerhard, as he called him, very well. He told them that Siebenreicher had actually lived at that address in 1932. "But then the house burned and doesn't exist any longer. What a sad tale! The fellow committed suicide by jumping out the window." So that was the official version of his death! Hans König went back to the town hall with the new information and verified that, in 1932, a Gerd Siebenreicher had lived at that address.

In a later recording, H. O. König heard, **"Gerd Siebenreicher didn't commit suicide."** Was the messenger referring to the conversation in the old pub? Gerd Siebenreicher hadn't committed suicide **"but was pushed out the window,"** as H. O. König learned in a later message. But after examining all the details and facts, the researcher never again heard from Gerd Siebenreicher. Why not? Did this spirit being contact the researcher only to free himself from a burden: the widespread allegation that he had committed suicide? We can only assume. In our connections with the Spirit World, much remains unknown.

After this crucial experience, Hans Otto König began to doubt his first working hypothesis related to the unconscious: "The animistic theory of paranormal phenomena, in this case, is sent into a tailspin. How could my own unconscious produce such a dialogue with all this detailed information that's unknown to me? There can certainly be no question either of psychokinesis or of what is called in parapsychology "the Case of Philippe," an experiment in England where a group of people purposely created a phantom only by the power of their thoughts."

Many other persons, totally unknown to the researcher and having lived in the most diverse countries, presented themselves to him during subsequent contacts. He always checked their identities. All had indeed lived!

"It takes amazing acrobatics of the mind," says H. O. König, "to be able to explain the power in my own unconscious that can produce these messages sent by people I've never heard of, but who actually lived. Some (colleagues) told me that the morphogenetic fields of Rupert Sheldrake could provide a theoretical explanation. I talked with Rupert Sheldrake a long time about it, having great respect for him, especially for his fantastic animal research."

But apart from the fact that the theory of the morphogenetic fields is still only a working hypothesis not yet verified by concrete results, I had to exclude it in my subsequent experiments, such as the "artificial head stereophony," in which you are able to locate voices. I received some very interesting messages like, **"I am now at your right hand"** (message heard in the earpiece of my left ear), or **"I am behind you"** or **"I'm above you."** These tests showed that a good selection takes place in the fields of information, and that these are not our thoughts that simply attract certain information of the surrounding morphogenetic field. In addition, if such were the case, during trials with several persons where each one was intently thinking about a different loved one, there would have been a total information-chaos on the tape. However, messages are always addressed and clearly directed to a particular person.

Another finding was also very important for the researcher. The "messengers" often respond directly to a specific question, posed verbally. It is, therefore, a particular phenomenon where you can observe a thought transmission and where the thought of the experimenter is known. As the messengers have confirmed,

"Listen: We want to say to the humans. We know all your thoughts. All your thoughts are known to us."

The thought itself is even more important than the question posed verbally, because during many recordings they answered his question before he asked it aloud. The researcher cannot count the number of times the communicating entities told him that they hear our thoughts and questions, capture them and know them.

"We hear your thoughts."

"We are aware of all your thoughts and questions."

Hans Otto König knows for certain that these messages do not come from the subconscious of the experimenter: "The one who actually occupied his mind with the phenomenon knows that this hypothesis must be excluded. Faced with the reality of the very clear dialogues obtained by my technique developed later, this question becomes even derisive. Forty years of intensive research allow me to affirm: Our personal consciousness survives physical death. Too many clues demonstrate this fact. In addition, I do not communicate with the dead or the deceased, but with pure and subtle "Information Fields" that exist in Universes parallel to ours.

The term "speaking with the dead" is, of course, and erroneous expression, because it is obvious, that we can never establish communication with something dead. And yet, this expression is used by the communicating entities themselves:

"The dead are calling for information."

"The dead do not come today no more."

"Although dead he now is calling."

Thus this answer to the question, posed by H.O.K.,

H. O. K.: "Where are you now?"

"Among dead friends."

This turn of phrase reveals another typical reality of the Instrumental TransCommunication: the Spirit World deliberately use words and signals corresponding to our reality and our way of thinking so that we can understand them well:

"We will communicate to you all terms and definitions according to your comprehension."

They very often add to their messages, **"as you say."**

But then what is death, or rather what goes on living after the physical death? Because "something" or "someone" goes through the door of death that opens onto other realities: in the data archives of the voices obtained by conventional methods, H. O. König received thousands of messages that communicate to us that death is not the end of existence. Some examples:

"I am dead. I am alive. I'm with König."

"Paul is arrived."

"Leni Schade is greeting, lives here with many friends."

"Frank Tölke is calling."

"I have a beautiful world."

"The dead speak; it's us."

"We are here, and we live."

"The dead are calling in thoughts and speak."

Then, of course, the very interesting and eloquent archives of all the other experimenters worldwide confirm these data; for example, the recordings by Monique Simonet and J. Blanc-Garin in France and by M. Bacci in Italy, the messages received by A. Cardoso in Spain, and the contacting by C. Nunes and S. Rinaldi in Brazil.

The objective reality of these measurable acoustic documents, the pertinent and appropriate answers to the posed questions, the possible identification of the interlocutor during his lifetime thanks to his transmitted information, the responses referring to the given situation, the references to particular places and precise details, the precognitive responses, the technical information, the fact that so many different messengers keep saying they are still alive, and finally, the spectral analysis technology—everything confirms that our personal consciousness continues to exist after our earthly death in worlds beyond our five senses.

One of the major problems of this research, however, lies in the use of adequate words to describe the encountered realities, the characteristics of which we do not know and cannot yet define. In this book, you will find the following terms: "psychic structures," "mind-structures," "information fields," "communicating entities," "spirit," "personal consciousness," and "souls," that all refer to the same reality, whose essence is still unknown. Throughout history, all these concepts have already received various meanings.

In the research of H. O. König, personal consciousness refers to the sum total of all personal experiences of an individual, which are always subjective. These experiences influence and form the mind of the person, who, because of them, possesses a certain "psychic structure," much like a spectral image. That structure remains alive even after physical death. But over time it will change with the new experiences that this "consciousness structure," this "thought pattern," this "entity" or "SPIRIT/MIND" will have in his new life.

But this description is provisional and not a definition. It's still too soon for that. On the other hand, with the countless studies and concrete results by all researchers — paying tribute to the works of Pim van Lommel, Sheldrake, Capra, and even quantum physics — one day, new doors will open, leading us to an understanding of all these realities. At this stage, it is certainly too early to define them, but it is indeed possible to work and achieve results with these realities whose definition eludes us.

As he often remarks, smiling, "Does science do anything else when working with gravity or matter? Can they really define gravity or matter? Or do they simply have a knowledge of matter? Similarly, it is impossible for me to define the 'information fields' or 'personal consciousness' with which I work. But I have many concrete results to show and have developed many devices that are capable of capturing these 'Information Fields.' And they keep assuring us that our personal consciousness survives physical death."

So the researcher leaves it to others to discuss these issues. To move forward in the research of these Unknown Worlds, he himself plans to put all his energy into the experience of that reality by performing multiple, extended trials: making observations without prejudice, analyzing them, reflecting on each tiny phenomenon and its minimal results, creating new hypotheses to verify them by other studies and concrete results, and rejecting them or not. And through this, he will try to achieve a deeper understanding of this reality, which is certainly not reducible to what is scientifically quantifiable and reproducible.

The discovery of the continuation of life after physical death is so essential for us humans, maybe it is too great for us to dare to actually look at it in depth without necessarily looking at it from a religious point of view. The stakes are enormous, both for the foundation of a "scientific" approach and for our own worldview. Criticisms are therefore tough and bitter. They range from auditory projection, pathological obsession, mediumistic psychosis, delirium, pure fantasy, a pact with the devil, and ridicule to the total denial of the existence of paranormal voices.

The difficulty we encounter when listening to the tapes tells some unbelievers that the experimenter himself is interpreting something in the radio sounds. It is true that many messages are very difficult to decipher, and there is a real danger that we might project the words we want to hear. The "Luxembourg effect," as we say in Germany, refers to the fact that words can be interpreted from hearing the first syllables, such as in well-known song titles. But the fact remains that the criticism is in no way grounded in the many clear messages received by so many people throughout the world for decades.

Another criticism of the reality of the phenomenon is that, because the short wave method is used, the voices heard are merely interference of other radio voices. Interference exists, of course, but is only a tiny part of the entire acoustic spectrum of voices. In the 1970s Konstantin Raudive made studies based on that interference problem, and more recently the same studies were repeated and broadened by Anabela Cardoso. Magnificent results put an end to the criticism about interferences once and for all and proved the paranormal character of the voices. Their experiments were conducted under the strictest laboratory conditions and the most rigorous supervision. In this regard, the researcher loves to listen to the two messages received from his mother, who was an opera singer during her lifetime. We must imagine hearing a little excerpt from an operatic aria:

"There is, Mrs. König here today,"

or

"Do you hear me Otto; do you hear Josef König?"

Mrs. König is the name of the researcher's mother, and Josef König is his father's name. Where should there be interference from another station?

When listening to the clear voices received by the devices the researcher developed later, some argued that H. O. König must have rigged them, precisely because they sounded too similar to the human voice: it was impossible to conceive that they came from a World Unseen. These people didn't want to know anything about the enormous amount of technical work invested in developing small bits of sentences into whole sentences, mini-dialogues into long passages and then long dialogues of incredible acoustic quality. Too many fears seem to hold us back. So it is better to confine the matter to the realm of faith, or to find even more unlikely explanations than to take as the point of departure of all research what is most probable: physical death does not mean the end of our existence. Therefore, paranormal voices are objective, measurable, and analyzable validations.

A New Adventure Announcing Itself:
The Beginning of the Fundamental Research

For our experimenter, all these observations gathered at this stage showed clearly that he had to review and consider the world in a different way. His own vision was toppled. Concrete results indicated to him the direction and the new way to go: try to improve the contact quality so more information is offered to him. It was the only option he had to better comprehend this phenomenon, which could shed more light on themes that the human being has been questioning for so long.

"I'd just tested all conventional methods of the ITC," said H. O. König. "These could not give me further information because the results were too fragmented or too sparse. That wouldn't get me anywhere, condemning me to remain at the level of hypotheses or assumptions. So either I had to set about working rigorously, laboriously searching for a new technique to achieve vastly better results and observing the new knowledge this brought, or abandon the whole thing!"

If the survival of our personal consciousness is a reality, he had to develop new technical devices that could receive longer and better transmissions from the Parallel Worlds. But what kind of equipment did the World Unseen need to transmit to him? By what technical process could the transformation of information fields of other dimensions occur? Furthermore, how could he stabilize the synthetically produced support in a way that would improve acoustic quality and contact duration?

At this point, the researcher definitively left the thousands of recordings made by all the renowned pioneers using conventional methods to walk the path of extraordinary basic research that no one had conducted before. There were, of course, some tentative attempts: for example, G. Meek's Spiricom and Seidl's Psychophone. But, according to the researcher, the valuable and honorable work of these two men didn't provide a new basis for a better understanding of the phenomenon:

"George Meek came to visit me in my laboratory in 1987. He told me about his device, Spiricom, and how it worked. He said he managed to get a long recording, and then nothing. So, for me, the device didn't seem very interesting because the tests were not "reproducible," the sound quality was weak, and, above all, there was no foundation we could build on. Seidl has certainly had the good idea to propose gathering frequencies with his "Psycho-phone." But this idea is somehow a mere extension of the radio method and doesn't shed any further light on the question of background research."

And on what is the background research by H. O. König based? "It is based on four essential pillars: low frequency (LF) and high frequency (HF) physics, a science of mind or mental physics, ethics, and, of course, as indicated by the term, a foundation on which we can systematically build, develop, and improve new devices."

While traveling this creative path, we must be open to the new, daring to leave safe territory and walk alone against all. So, with the absolute support of both his late wife Margaret and his son, Markus König, and in the company of all "those" that our eyes and ears are normally unable to perceive, he set sail on "a scientific journey" of another order.

The Laboratory

Discovering Other Vibrational Worlds. A Search for Another Order

*Obstacles cannot crush me; every
obstacle yields to stern resolve.*

~ Leonardo da Vinci

By losing purity of the heart, science is lost.

~ N. Valois

Experimental research revolutionizing our common frame of reference

Nowadays, each branch of science is based on a theory that gives us a foundation on which we can build a model. ITC, however, represents "a scientific anomaly" that still lacks any theory on which to rely. Therefore, each researcher is obliged to base his research on the observation of his personal reality, on experiments, and on more and more experiments. He can only proceed empirically.

These are the words with which the researcher often opens his lectures. Listening to him speak is like listening to an archaeologist passionate about physics and the Spirit World who completes many small excavations and makes small technical discoveries, enabling him to advance step by step in his understanding of this mysterious acoustic phenomenon and in the creation of new technical devices. It is child's play for a man like him, a specialist in electrical engineering, to build a radio whose frequencies are well known, but to develop a new technique for contacting structures whose frequencies and partners are unknown is something else entirely!

Understanding the radio phenomenon is simple, but comprehending the phenomenon of ITC is much less so. H. O. König declares, "Normally there is a transmitter and a receiver working on the same frequency. But in ITC, two different physical worlds must come into contact. We are in a material dimension of space and time, and the others are in a non-material dimension beyond space and time. So the greatest difficulty is the temporal synchronization."

What to do, which direction to take, what technical process to use, and what to base it on? Hans König likes to recall a message received years later from the Spirit World: **"When you get to communicate with a drop of water, you get the answer to your questions."**

The macrocosm in the microcosm? The spirit in matter? The thorny debate about matter and spirit inevitably resurfaces in this research. What is matter, what are mind and consciousness? Are they compatible or incompatible? The scientists have willingly set aside the problem of consciousness, complicating everything. How can we speak about the breadth, depth, and height of consciousness or thought? When studying ITC, this is essential.

Instrumental TransCommunication is at the crossroads of the Physical and Psychic worlds: the Outside World (a technical process, an objective acoustic document with a certain content and spectral images that can be analyzed) and the Inner World (the composition of the Spirit, structures and information that one cannot measure, calculate, or define).

ITC, therefore, assumes that a non-physical reality exists. And here things get complicated for any rational mind. This research requires that the "structure of the thinking mind" of the researcher himself plays a critical role in the process and the results obtained. "It is clear," says H. O. König, "that the existence and importance of psychic structures, especially their permanence, undermine the very foundations of our conventional, materialistic science. Including supernatural structures in the scientific model would mean a radical rethinking of matter, and even the cosmos. The stakes are too high."

H. O. König is convinced that a scientific mind is above all capable of revealing new knowledge, and for this it opens without prejudice to the world around it and observes what reality conceals as phenomena. Here, acoustic documents with electronic voices, which tell us that they are coming from the dead, assure us that physical death does not mean the end of the existence of our personal consciousness, and testify to an interaction between different levels of existence. He wants to understand this phenomenon and see how it actually works.

The researcher argues, "First, a simple observation: we need a voice, writing, or another medium so the thoughts become a reality for others. Therefore, in what reality do the thoughts find their existence? The "information thoughts" don't seem to be integrated into the three-dimensional world, that of the five senses. Until now and to my knowledge there is no device capable of recording our thoughts and playing them back immediately. And yet our thoughts, before being expressed, do seem a reality for the one who is thinking them. If the reality of supernatural structures does exist in another dimension—the fifth or sixth, as Professor B. Heim postulated—it is a question of transforming them so they can be heard and become a reality in our three-dimensional world!"

All his research, among other things, revolves around a single issue: the transformation of "invisible information fields" that exist in universes parallel to ours. Finding answers to this question is like going in search of a new world: the road is not yet mapped out, the land is still unknown, and the success of the venture is still doubtful. During this exploration, we often grope in the dark. Then we see the first small glimmers of hope, calling them intuitions, reflections, and dreams—tiny messages received by radio and multiple failures. The researcher adopts Edison's wise view: "I am not discouraged because every unsuccessful attempt left behind is another step forward."

This is the path of experience, a relentless day-by-day dedication that allowed the German researcher to build new systems of contact, to improve and qualify them over the years, and to present the results to others.

Opening these massive technical trials to the public was another central pillar in his 40 years of pioneering work. He wanted to give everyone the possibility of a personal experience: "If a person claims to have succeeded in such an experience at home, that person is obliged somehow to believe or not believe. Live contacting trials, on the other hand, allow each participant to reach his own realization. To believe in something, we can get knowledge through essays, or we can get information through direct experience. The difference is crucial. That's why I always conducted my big tests in the presence of an audience, and I agreed to do several live experiences on television or radio programs."

This experimental research exploring new worlds requires, of course, other paradigms, but even though such research belongs to another order, it does not relinquish objective facts, measurable or in some way reproducible. It takes reasoning and critical thinking. The work raises big questions because it lies at the crossroads of disciplines that normally do not meet: *there are words like "humility," "love," "honesty," "knowledge," usually reserved for the fields of religion, ethics, spirituality, or even esoterica.*

Love, knowledge, spiritual evolution: by what criteria can we define or analyze them? These value systems, so difficult to capture, especially by materialistic and scientific language, are essential realities for those wishing to get in touch with the Spirit World. Their meanings will be revealed and opened slowly only by personal experience, and the many moments spent in the company of entities from other realms. But they are the *sine qua non (the essential condition)* that enables us to journey to other universes. In doing so, we are confronted with unknown realities, even inventing new words.

Research in Collaboration with the Spirit World

Without being open-minded about the unknowns of other realities, without the inner attitude that is necessary for making contact trials, any research in this field will have no results. As difficult as it is for a rational mind to accept, this kind of research, always linked to the experimenter, can *never succeed without the help of the Spirit World.* This new technique was not created collaboratively by a team of specialists, inspiring each other in a laboratory and financed by the government or patrons.

On the contrary, the new technique was conducted independently, but in continuous alliance with the Invisible World. That inspired and enlightened the researcher, and gradually transmitted the data needed to increase his understanding of this astonishing phenomenon. As the researcher says, "How could it be otherwise? They alone really know how they affect the technique, and the frequencies needed for contact. Without their help, it would be like searching for a needle in a haystack."

Here are some examples of messages received:

"We give information for the technique of Hans König."

"We give you help and Information for your re-search work."

"The stabilization is not good at the moment. Try to change it."

"It's better now."

"In the Zentrale, many scientists have set up them-selves to help in improving the contact bridge."

"Listen: Hans König received information from us to the possibility of contacting."

Thanks to the suggestions of the Beyond and a meticulous, gigantic body of work, the researcher was able to leave a technique based on the five senses and move on to a new technique beyond that.

The Spirit World Often Communicates with Simple Images

Most of the time, the clues received from the spirit world are symbolic. It is up to the researcher to transform the given image data into practical technical information to use for his initial contacts. Sometimes he also received precise technical data. For the In-frared System and the development of the HRS-System, six frequencies were communicated to him.

But these frequencies were revealed to him only after he himself discovered them and their significance after searching for one year in his laboratory. "You never get a ready-made recipe," he admits. "On the contrary, I often felt I had been put to the test, as if the Spirit World wanted to reassure me about my true intentions and the depth of my interest in my research work. It's like they're watching me before revealing important information."

The following message has been communicated repeatedly:

"Hans König, we're watching you."

During a conference in Wesel (Germany) in 2004, he stated: "It was in Bad Kissingen, during a contact, when I asked if they could communicate to me something important for my research. I received the following message:

"You need a mirror for our image."

"If this sentence seemed insignificant at first, it wouldn't leave me in peace during the night. I thought about the image of the "mirror," its meaning and its implications, in technical language."

Another time, when I asked what I should do to improve the contacts, I was told: **"Parabolic-dish."** Now it was safe for me to assume that the notion of the mirror was very important, and the concept of the parabola taught me the need to focus the proposed frequencies. So for some time, I made trials with a parabolic mirror, unfortunately without results. You see, information is given only gradually, and always in images, which for a scientific mind is not the common language.

"We must be ready to abandon sophisticated mathematical formulas and go to simpler interpretations. The "parabolic-mirror" is an image of startling simplicity, but it is very important if you think about it. It gives you a mental picture for constructing a new technique other than the existing technique, a procedure known by any physicist. I can only repeat: "According to my own knowledge of physics, such contacts seem impossible."

Once, the world of the Spirit gave him information about the electronic equipment to use, but it was already so old it no longer existed on the market. Once again, he had to readjust and rethink the received information. Personally, I (Anna Maria) had the chance to be present when he received six frequencies for the Spirit World that he could use with the UDS-System.

"Contact panel closed to Anna Maria Wauters and Hans Otto König."

"And now the frequencies: 930.7 / 931.7 / 930.42 / 930.8 / 931.2 / 930.9 nanometers."

Another day he received, **"Change the phase for 20% percent."** After changing the phase, we heard, **"It's better this way."**

Yet, as some have noted, the transmitted information was poor. There was no detailed technical information. So the afterlife will not or cannot tell us about new scientific and philosophical topics "unknown to us," or do those exceed the "average level" of the auditors? What does this man think, who has so often frequented the Worlds Unseen:

"I think the way the spiritual entities are proceeding is very wise and just. If they were sending us technical information, like an accurate mapping or a general recipe, it would be too simple. We would then immediately be transported to their realm without effort, without personal commitment, without deep reflection. It would, in fact, teach us absolutely nothing, and we wouldn't need to walk the path on our own because they would immediately give us their knowledge, which is in some minds obviously very broad and not commensurate with ours. In fact, they would take from us the joy of discovery, the adventure of research, and multiple tests, but most of all the gradual and profound understanding of the phenomenon, the mystery called 'Life.'"

There may be *savoir* (a knowing, or appropriation of information), not *connaissance* (knowledge more intimate and personal), not gradual penetration into the depths of the mystery. Besides, what would we do with our ample knowledge passed entirely out of proportion with our own brains and maturity? Are we so sure that humans would know how to use the ready-made recipes that they would send us? They do not convey brilliant intellectual knowledge: the maturity of mind is crucial.

The Spirit World Communicates in Dreams

Since childhood, the man was amazed by his ability to receive messages in dreams. The time we spend sleeping seems favorable to communication with the Spirit World, with a loved one or a spirit guide who enlightens us in the darkness of the night. Platonists were already convinced that, during sleep, our soul was freed from the body and able to wander and receive messages, insights, and lessons. H. O. König often got new opportunities for his groundwork, new trails to follow, and inspirations during dreaming.

Also, he was often asked to pay attention to his own dreams, receiving messages like:

"Take care of your dreams!"

H.O.K: "Which crystal should I use?"

"Berg crystal. Have a little care! Have spoken in your dream."

Thanks to those subtle images, the researcher received the most important vision during his entire research!

Several nights in a row, he had the same dream: An old, bearded man stood in front of a table hidden by a kind of curtain. He opened the curtain and showed H. O. König a card on which was drawn a schematic representation of a technique. This proved to be the draft of the Multi-Oscillation System, the basis on which all subsequent devices were built. If the researcher had considered this schematic draft just like any other dream, nothing special would ever have happened in his research. But he trusted it, beyond rational thought. A dozen years later, the same bearded man manifested on a television monitor, a visual contact made in 1995 at Büdingen (Germany), with the following statements:

The Spiritual Guide of H.O.K.

Now you hear my thoughts.

What you see, is one of the many bodies that we can possess.

Few people know me.

We have great concern for what is happening on your Earth.

Recognize that we are in connection.

We're reiterating, connect with the Universe.

The Universe is all that is alive.

Recognize in the legislation.

For this, you must learn."

And do not follow all the things offered to you and you do not understand.

You have to learn; get to know yourself.

You learned; the crystals are of great help.

We know that extraterrestrial life will come to get in touch with you.

The search to get in touch with us by TV is making great strides.

Hans König knows the goal. He'll reach the goal.

We pass a lot to humanity.

We thank Hans König for the opportunities and enormous work for research.

We have lived many times, crossed the whole evolution.

My energy was also in a crystal.

The Course of the Basic Research

The course of research, the stepwise procedure that leads to the creation of new devices, requires a sophisticated and refined knowledge in electronics. It requires a specialized technical presentation that is not appropriate in the context of this book, and for which I (Anna Maria) am unable to find the exact technical wording. Paradoxically, this journey of discoveries at the technical level is the "basic core" of the research. The daily laboratory work, experimenting to infinity, is the essence of the researcher's life! Technical research of this magnitude is still unique in the world of the ITC. As he puts it:

> I am still waiting for a colleague with whom to share and freely discuss our mutual technical breakthroughs. So far, the information was, to my huge regret, always one way. This may sound arrogant, but it is a fact. I have had many visitors in my laboratory, but none gave me a lead to a new technical discovery or any suggestion on the subject. Among them, there

were many well-known people, who had long paranormal dialogues, even by phone, apparently always
completely spontaneously. But they never invited
me to any of their trials, and they have never been
able to explain to me any of their procedures.

In short, they all asked that I take their word for it.
My way of working is to check everything and try to
understand it, even if it exceeds current scientific
thinking. There was no shortage either of physicists
who offered me theoretical recommendations for im-
provement, but none of them were ready to get to
work and have paranormal dialogues by themselves.
There were also people who tried to copy my devices, but this is not so simple because each researcher has to forge his own way.

Unlike other electronic discoveries such as radio, which are
infinitely reproducible by a technical process independent of the
inventor, the development of devices for "bridging a contact" to
"Spiritual Structures" is closely related to the experimenter himself.
The Multi-Oscillation System, the Infrared System, the HRS- and
UDS-Systems cannot be reproduced using the plans of the researcher with good results. His plans are, admittedly, a good basis
for work. They are a sort of map, a guide, or, to use an image that
even the researcher would never use, like the famous Compostela
shells marking the route. But we must find the path ourselves. We
learn to know ourselves and to know one another, to discover our
own frequencies, then to open them so we can receive the entities
coming from the Spirit World.

The Two Basic Principles of the "Contact Bridge"

The researcher often said, "In fact, everything is very simple. It is we who make things extremely complicated with all our
constructions of thought. For something to be accepted, it must look
complicated. If it is very simple, most humans reject it in general by

saying this is not possible, it's too simple. And yet the whole process of the Instrumental TransCommunication is very easy, but at the same time very complex and difficult to achieve." Making contact with the Invisible is based on two fundamental pillars: The principle of resonance and the principle of "All is One," connecting all.

"We are able to establish communication with subtler worlds for the simple reason that they're connected with us. This is remarkably simple, but so complex in its implications for reason and for human life."

The results of ITC demonstrate that parallel worlds related to ours do exist. A large system in which all visible and Invisible Worlds are integrated is governed by the principle of resonance. The latter is critical for the phenomenon of paranormal voices. We can only get in touch with entities who are, to put it simply, on the same wavelength with us. The following experience of the researcher is a good example of that principle:

One day he turned on his multi-oscillation system and an entity named Kurt (out of respect for the family we include neither the name nor the exact address transmitted through the system), who asked him to tell his wife that he is still alive. The researcher was not particularly happy to respond to this request. He didn't have the time or the desire to go running up hill and down dale to find a completely unknown person to tell her that her husband had asked him to say he was still alive. He had known people shipped off to a mental hospital for doing less.

He replied to him, "Frankly, there are so many people who make contact. Why did you choose me and present yourself to me?" Kurt's response was simple: "**Because we are of the same vibratory level.**" (Kurt used the word "co-vibrating.")

The investigator, however, replied, "It's not so simple. If you want me to visit your wife, maybe you could give me a more convincing sign to let her know that this message came from you."

"Is busy preparing the wedding of our daughter."

The researcher didn't consider that a really convincing sign, but he nevertheless took his tape recorder and set off in his car for Düsseldorf to the provided address. When he arrived at the house, he rang the doorbell, and a very old woman opened it. He introduced himself, but she took him for a salesman and shut the door in his face. He rang the doorbell again. This time it was a young woman who faced him. He knew that what he was going to tell her couldn't be more strange, but he had promised her husband, Kurt.

The woman, however, replied, "I'm sorry, sir, but my husband is dead, a long time ago!"

"Yes, I know, ma'am, but he asked me to tell you that at this moment you're preparing for your daughter's wedding."

A long silence followed, then the woman took him into the house. There, H. O. König let her hear the recorded short dialogue with the entity that had presented itself as her late husband. The researcher learned that the couple had often had long conversations about the great transition and the possibility of life after death. Shortly before her husband passed to the other side, he had made a solemn promise to his wife: If an afterlife did really exist, he would try to do everything to let her know.

Was this promise so engraved in his consciousness that he did everything to fulfill it? And how did Kurt know that the researcher was a man who would be true to his word? The woman was very grateful to receive this message from the man she had loved so much and with whom she had been so deeply connected during so many years, and even beyond. H. O. König took his leave, reflecting on the implications of everything he had just experienced. At home, he went back to work. His observations then allow him to affirm today:

> The concept of Instrumental TransCommunication consists of a triangle: an experimenter who is at the same time both a transmitter and a receiver—from a technical device and a communicating entity that is also transmitter and receiver. This triangle, this trin-

ity, should really be in complete harmony so a contact can be established. Without resonance among the three parts, nothing will work. They are like three cogwheels in a clock, where all the little teeth must mesh completely for the clock to run.

This harmonization and stabilization are of the highest precision, of superior subtlety and extreme sensitivity. It requires a particular passion for the technique, as the researcher notes: "Spirit and matter form a whole. I think, for example, of the technical apparatus as a kind of living being with which I connect, strange as it may seem. There are many people who believe the technical apparatus is dead matter. Nothing is less true in my experience, because when I develop and create something, I give the thing a spirit. And the spirit and the technical apparatus, connected as a whole, become one.

The Invention of the Multi-Oscillation System

The first device was named by the German press "Der Generator" (The Generator). Even if the term was technically incorrect, from a media standpoint the name was dead on, because in the 1980s, this device caused a groundswell of sensation and brought in thousands of letters. Thanks to this new system, for the first time we could hear, during a public, live transmission, more than snatches of words or whispers, but actual paranormal voices speaking. It was a short dialogue whose sound quality was far superior to anything we had heard before. This was the incredible result of four long years of laborious work.

The foremost question of this electroacoustic phenomenon with, on the one hand, an acoustic component and, on the other hand, an electronic component that makes the former audible, is how the information is supported by the Universe of Spirit. What precise medium do the energy beings use to modulate their thoughts into speech? Finding an answer to this was the first major challenge of H. O. König's research:

The Multi-Oscillation System is the result of a logical development of electronic voices received by many experimenters around the world. Making recordings only with the radio method made it nearly impossible to find a pattern because the interferences were too brief and too mutable to be analyzed. On the other hand, among the records made by using rippling water as background noise, I noticed that sometimes longer messages came through, short sentences that indicated a more consistent stabilization. Maybe I could find answers if I did further investigation in that direction. Then one night I had the idea to analyze the frequency spectrum of water, realizing that some were found in the Ultrasound Field between 30 KHz and 80 KHz, imperceptible to our ears. Maybe these were a starting point for a transmission device that could record voices coming from psychic structures.

He began to separate these frequencies from the others and to create new ones synthetically with oscillators. The results improved immediately. Isn't it striking that the discovery of this technical system is expressed in biblical imagery: "The Spirit of God moved upon the face of the waters." And isn't it surprising that American Indians often sit in a river to connect with their ancestors? Although running water has a character similar to spoken language, and the risk of projection is, therefore, great, the fact remains that the language of "psychic structures" is detached from it. A trained ear can hear the difference.

Six significant frequencies

But with the joy of discovery came new challenges: how do you ensure that different entities can manifest during the same contact, and how do you get a broad spectrum of frequencies?

After hundreds of daily experiments, he discerned that a combination of six frequencies was significant for contacting the Invisible World. If he removed only one of the six frequencies, the contact stopped. This "Six-Frequency Combination," so to speak, always manifested parallel to the positions of paranormal messages, clearly visible on the spectral images for anyone able to decode a radiometer. But the combination of frequencies always changed according to the communicating entity. Each "Information Field" had its spectral image, or diagram, quite different from others.

After developing a system that allowed him to measure frequencies of a living person, the researcher noticed that every being seems to possess a unique image of six frequencies. He made this analysis with more than twenty people, and each one had a different schematic. Was this "spectral thumbprint" of the living person the same as the "Six-Frequency Combination" of the being who has passed on to another life plane?" This was the essential question.

One of the participants in that project was a female doctor, seriously ill with leukemia. She asked the researcher if she could participate in his research. When the time came, she insisted that he set up his technical equipment in her hospital room. She didn't want emotional consolation, she said, but just to participate until the last minute in the adventure of human consciousness. The researcher hesitated, but she assured him that she wanted to do it at all costs.

The goal of their joint experiment: Would her frequency image, measured when she was still alive, reappear on the Oscillogram after her physical death during the contact and, if so, for how long?

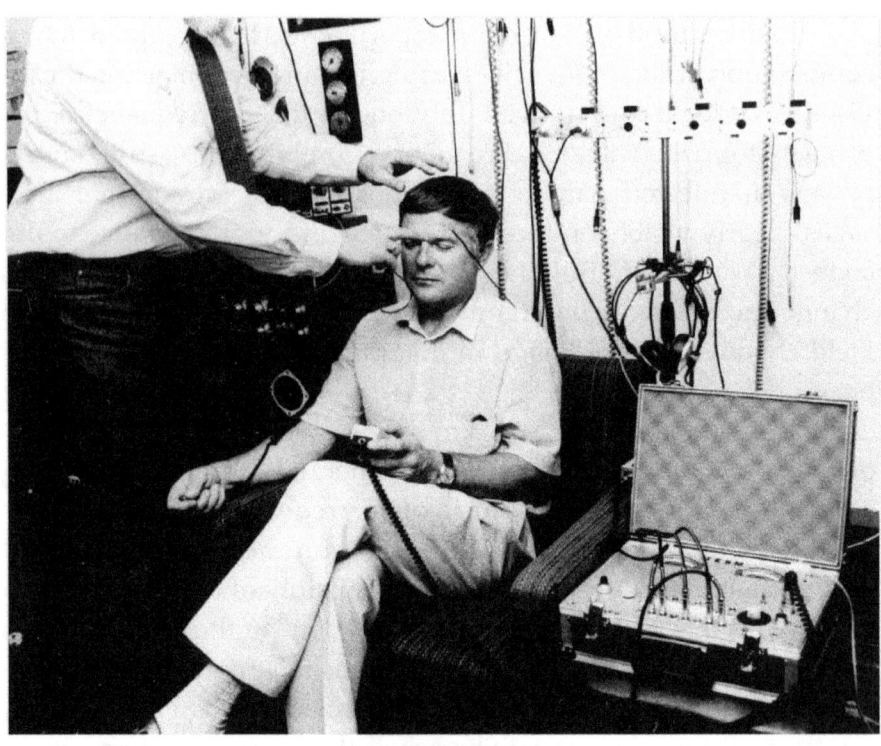

Measuring the Frequencies

H. O. König and the chief physician of the private hospital were present at the great transition of that wonderful woman. They waited. Fifteen minutes after clinical death, H. O. König received her waveform image on his oscillogram, containing the same "Six-Frequency Combination" he had measured in her lifetime, with the following message: **"I see my body."**

The chief physician was so overwhelmed by what he experienced in that patient's room that he spoke to his medical team, proposing, in collaboration with H. O. König, to develop a much broader project, including as many test persons as possible, to see if the results would verify those seen with the female doctor. In the months after he made his proposal, this competent chief physician was dismissed from his post. He was no longer credible among these other scientists. There was nothing left to do but settle into a private medical practice. Nevertheless, H. O. König wanted to continue these test studies that *could provide convincing proof on the permanence of individual consciousness after physical death*. But such projects, unfortunately, may not resonate in the powerful research institutions of today's society.

But let's go back to the laboratory of the researcher, looking into the following problem: How do you propose a spectrum of frequencies large enough for different entities to come through? It was not clear at all, and he had no idea. He searched for months but could not find an ideal solution. Then for several nights in a row, he had the same dream (described previously). An elderly man showed him a schematic image, "A Block Wiring Diagram." He began to comprehend it, but still questioned and hesitated. All in all, he was only following his dreams. The results, however, confirm the unorthodox approach of the researcher.

But the third major technical barrier arises: How do you stabilize the oscillators? To remedy this, he made the following test: use six pieces of quartz integrated into the six oscillators. Because he could not find clear quartz on the market, he needed nearly six months to adapt them manually: he polished and refined each quartz, layer by layer, rebuilding them in the box so he could check if the right frequency was reached, then removing them again and repeating this process as often as necessary to reach the best frequency in each of the six oscillators. After four years of hard work and much patience, he managed to make the first contact with his Multi-Oscillation System.

He will never forget the first words he heard:

"Leave on. Look, we are alive."

The following contact was made in 1981 at Radio Luxembourg, during Rainer Holbe's program *Unglaubliche Geschichten (Incredible Stories)*:

H.O.K.: "Can I try to get in touch with you?

R: " Try."

H.O.K.: "Can you hear me? I think I have the right frequency.

R: "We hear your voice."

H. O. König is now addressing a deceased friend named Helmut.

R.: "I come to Fulda."

(Fulda was, at that time, the conference venue.)

Then another voice said, **"Otto König speaks with the dead. (Totenfunk) "**

Totenfunk is a German neologism, which literally translated means "dead-radioing." This sentence, which does not lack humor, was broadcast throughout the day on all German radio channels in diction typical to the Multi-Oscillation System, somewhat technical and choppy, like on a computer.

Thanks to this new system, Hans Otto König made the acquaintance of people who would accompany him throughout his research: **the Central (die Zentrale).**

During one of their first contacts, the researcher asked:

H.O.K: "Who are you?"

"We have no names."

 H.O.K: "Do you have an important function, Zentrale?"

"Yes."

Since then Hans Otto and Margaret König have used the name "Die Zentrale" for these Spiritual Entities still manifesting themselves in the first person plural, recognizable by their spectral image. These entities say that they come from the highest cosmic level. They have always given clear and coherent information. They leave us entirely free, while teaching us that the time has come to change our vision of and our attitude about life.

They often begin their messages with **"Listen,"** and they do not expend more words than necessary. They don't have a sure formula for life, but give us deep insights to see the Earth with different eyes, a connection with the Universe where **All is One.** Personal occupations and personal stories of life aren't relevant for them, but they show great interest in the existence and construction of a bridge of authentic and true contact between human beings and the Spirit World. That is why at a particular moment in 1990, for example, it was important for them to announce,

"If other groups say they're having contacts with the Zentrale, then it's all wrong. It is a matter of pseudo-contacts."

At the request of the famous Luxembourg television station RTL, Hans König conducted many live experiments with the multi-oscillation system on radio and TV. He knew that the conditions of making contact in these circumstances could not be more difficult, and were even contrary to the inner concentration indispensable to the exchanges. He therefore always consulted his partners from the other realm before accepting any request whatsoever. It was essential to him that they not only agree, but also that they work together during the transmission.

Back then, it was always the Invisible World that prompted him to accept these proposals. They reassured him, promising that a connection would certainly take place during the transmissions, and they always kept their word. But nowadays the Invisible World has made it quite clear to him that the contact will no longer be established in the presence of persons coming only for sheer sensation or to satisfy their personal and intellectual curiosity. H. O. König no longer accepts television appearances with his new devices, where the contacts are incredibly clear.

Among all the transmissions, one stands out. It was made in 1986 with Mrs. Tölke and demonstrates that his first theory of voices coming from the unconscious is very difficult to sustain in this work. Frau Tölke had lost her son Franck in a tragic accident. Through many detours and "multiple hazards," she finally found her way to the König family, where she became aware of the possibility of connecting with the Invisible World. She set herself to work and established multiple contacts with the radio method. She became a valuable contributor in the FGT and received beautiful messages.

H. O. König installed the Multi-Oscillation System in the studio after many checks and controls made in advance by specialists and technicians of RTL. Mrs. Tölke, host Rainer Holbe, and of course the researcher himself were present on the set. He turned on his devices. The live experimentation could begin. Mrs. Tölke asked her deceased son, "Franck, are you there? Can you manifest?"

At that precise moment, a photo of the son was projected on the television screen. Instantly the live message came through:

"Now you can see me on TV."

Rainer Holbe informed the two experimenters that a photo of the son had just appeared on the screen. The two of them obviously had not seen the photo nor could have known about it. In that case, how could that response be a projection of the subconscious of the experimenter?

Then Mrs. Tölke asked her son, "Can you give us your name, so I know that it is you who has spoken just now?"

"Franck greets the mother, hello Dinchen."

That was the answer that so many viewers heard directly. **"Dinchen"** was the nickname the boy always used to call his mother during his lifetime. Then he added, **"Franckieboy"**—the nickname his mother used to call her son.

Dr. Gruber, a parapsychologist, commented about the show: "Experiencing and hearing live something extraordinary like this in a direct broadcast, that has never happened before! The research of paranormal voices during the last years could make huge developments, not least thanks to H.O.König's voice-generator."

With the development of this system, the experimenter had attained the first step on the path of his research: find a solid working base on which it was possible to develop other systems. When one day he asked his invisible friends how he could improve the system, he received the following answer: "**Come with red, infra-red.**" Then the Central transmitted to him six new frequencies in the 930 nanometer zone. The researcher didn't hesitate a second to put into practice the received information. After he had evolved two other devices, the FG 3 and the FG 5, permitting so many people at home to get quite better contacts with the afterlife, he prepared his Inner Self to embark on the discovery of another unknown, based on his rich experiences and his many discoveries.

The Infrared System

The researcher wondered how to create a new system using infrared rays, falling within spheres of light that cannot be perceived by our senses. Would the new system improve the quality of contacts? What new technical discoveries awaited him on his pioneering path? What problems would he encounter? And what treasures would he unearth from these mysterious worlds? Who would come to meet him, and who would or could be transmitted by this new device?

What will they converse about? Nothing could be more exciting for this man than to continue this adventure of the spirit in the company of his technique and his invisible friends. He already knew that the most difficult problem to solve would be again the stabilization of the diodes. He had the idea of incorporating very specific infrared diodes. Finding them was an adventure in itself. He had to try seventy diodes to find the six he needed.

Thanks to these, he built a consistent Infrared System that allowed him to project the frequencies transmitted on a receiver, a "picture element" that captured and then amplified them. This stabilization was not as prone to external influences as the Multi-Oscillation System. We won't go into the details of the development of that new device, which again required months of intensive work. But we'll illuminate some specifics.

First, the system worked well only in the presence of two experimenters: H. O. König and Marlene Dohrmann. The Invisible

World expressed that in a recording with the Generator in his laboratory:

"We need the energy of Marlene Dohrmann for stabilization of infrared contact and new knowledge."

Marlene Dohrmann was a member of the FGT and a very committed collaborator. She organized many conferences and was, for a long time, editor of the magazine, *Die Parastimme (The Paranormal Voice)*. She was very sensitive and had very good contacts with the Spirit World. The researcher tried to understand why the energy of two people was necessary for this system, but he didn't find any satisfactory response. So for many years H. O. König and Marlene Dohrmann worked together on all public experiences. There were many trials with this system, through which messages arrived of another order than those given by the "Generator."

The second feature of the system was that there were no pauses between the messages, so it was not a dialogue system. The words of the invisible came suddenly, and the pace was often extremely fast toward the end of the contact. Messages, however, corresponded to certain questions posed by the audience during the preparations for contact.

"The change to the Infrared System causes new possibilities for contact with us, and we will answer many of your questions."

Messages of a more philosophical nature were transmitted by this new system. They told us about our existence, life, death, contacting, and so much more. The Infrared System allowed us to reach higher levels in Parallel Worlds, as they informed us. Entities living on the fifth level were now able to pass through this new technique, although the station continued to introduce each test. Among these were a good many the researcher did not know, including many children, such as these examples:

"Listen: We tell you that we are fine. Stefan, Marco, Wolfgang, Birgit and Anja.

We greet our parents."

or

"Many beings on our side and connected with the Zentrale are here and greet you."

"Anja, Birgit, Wolfgang Frank, Heinz, Silke Semmel, Melchers, Claudia, Marko, Köchi."

"Many will come to show on TV."

"Listen. We are all alive and look forward to seeing you again."

Some Spirit entities seem able to descend from a higher level to lower levels. One account in particular talks about that. It said, **"Get to seek the dead from above, just touch."**

To the question, "What does 'above' mean" came the response, **"The higher planes."** How these superior minds can return to lower levels is still a great mystery for the researcher. They informed him that this process requires a lot of energy. H. O. König emphasizes, "These contacts are really a wonderful gift that invisible entities are offering to us. We must not abuse it. These communications are not to be taken for granted, so we must show sincere gratefulness."

Such tests require an interior preparation, a peace of mind and soul. Both Hans König and Marlene Dohrmann were doing daily meditations well in advance. The Zentrale sent the following message to the audience:

"Listen: For having a good contact with other planes of being, it is of utmost importance to prepare inwardly and spiritually."

The power of the mind, both of the communicating entities and the experimenters, is essential in the contact made with some more advanced levels of the Afterlife.

"Listen: Marlene Dohrmann and Hans König are sustaining us for the possibility of the transmission."

To celebrate communion and their presence among us, we put our best interior dress because the invisible is there when we are recording with the devices, often within our reach, but always audible:

"Listen: When we make contact, we are with you."

Then, a little later, they continue with the following words:

"Listen: Our words have a meaning for beings who understand."

This sentence is strangely similar to that of the Gospel: "He, who has ears, understands."

Through the Infrared System, the Invisible World also announced the next step in the research of Hans König: the transmissions of images on a monitor screen.

"Listen: In the near future, we will show the reality of our existence television."

But before addressing this technical, extremely complicated part, we must focus on a stage where the researcher developed a system based on this information:

"Try with a laser."

This message was sent to him in response to his question, "How can I succeed in having longer contacts?"

The System with Rubin-Laser

First, he had to decide what kind of laser to use. That's when he fell by chance on an article in a medical trade magazine talking about the Rubin laser. He had the idea of using this kind of laser "because its rays," as he says, "fall within the range of nanometers. They are very consistent, very powerful, and capable of stabilization over long distances."

Working with this material, he had to be very careful and use safety glasses. The Rubin laser was a good idea, because shortly after his decision to use it he received this message: "**The Rubin laser is the way.**"

He developed a whole complicated system with small special reflecting mirrors, as indicated by his working team on the other side with the following words: "**Deviation with special mirrors.**"

He ensured that the frequencies could initiate vibration on their own. He needed nearly two years to develop this technique, which was able to produce extraordinary stabilization and give the invisible entities a never-reached possibility for the modulation of their thoughts. Unfortunately, he encountered a mechanical problem, insurmountable at the time: this device with multiple mirrors was so sensitive that the smallest external vibration, as simple as a breath, affected the whole system.

He made several attempts to protect the system, but they all failed. He had no alternative but to invent a specific empty space where no outside influence could enter, but this theoretical idea was not feasible, and so he abandoned the project. He is still convinced that the device contained some fabulous opportunities for making contact. Meanwhile, he had already turned his attention to another idea where new challenges awaited him: the creation of a technique where invisible entities could manifest themselves by sound and image simultaneously.

The "TV-Generator"

In the early eighties Klaus Schreiber, a member of the researcher's association, FGT, phoned to tell him agitatedly that he had received images of his deceased daughter, Karin. Hans Otto and Margaret König had received many phone calls from people who claimed to receive paranormal images that, after thorough investigation, turned out to be normal photographs. So the investigator did not immediately respond to Schreiber's invitation.

Schreiber called him again, informing him that he had even received images of his deceased wife and grandmother, who appeared inexplicably on his television screen. This time the researcher immediately set out to visit him at Aix-la-Chapelle. After closely observing and studying the simple technique Klaus Schreiber used, H. O. König was convinced that the images were authentic, although the risk of projection was great with this process—for the researcher, too great. After a lengthy discussion, the researcher returned home, reflecting on what he had just experienced and remembering the many messages he himself had received on this question: Do you still have other options for transmitting yourself to us?

"Contacts with us in television."

or

"We have the ability to transmit our life with the contacts by television."

Immediately he contacted Rainer Holbe, the famous journalist and TV presenter, telling him about his encounter with Klaus Schreiber.

The meeting with this man, retired by now, was the catalyst for him to develop a new system of contact having both sound and image. This dual modulation system was completely different from Klaus Schreiber's and extremely complex technically. We could say that the level of the image and the tests, according to Schreiber, corresponded to the radio recording methods and technique of the researcher's system he had developed during his basic research. No other system required as much work as his tests with television, where it was necessary to synchronize sound and picture.

Thanks to his experiments undertaken between 1988 and 1996, real treasures were transmitted, like the marvelous text with the image of the bearded man we have already met. Other images appeared on the monitor screen, such as an emerging face and its gradual disappearance, to name only a few examples:

During the course of these images' appearing, the following words could be heard:

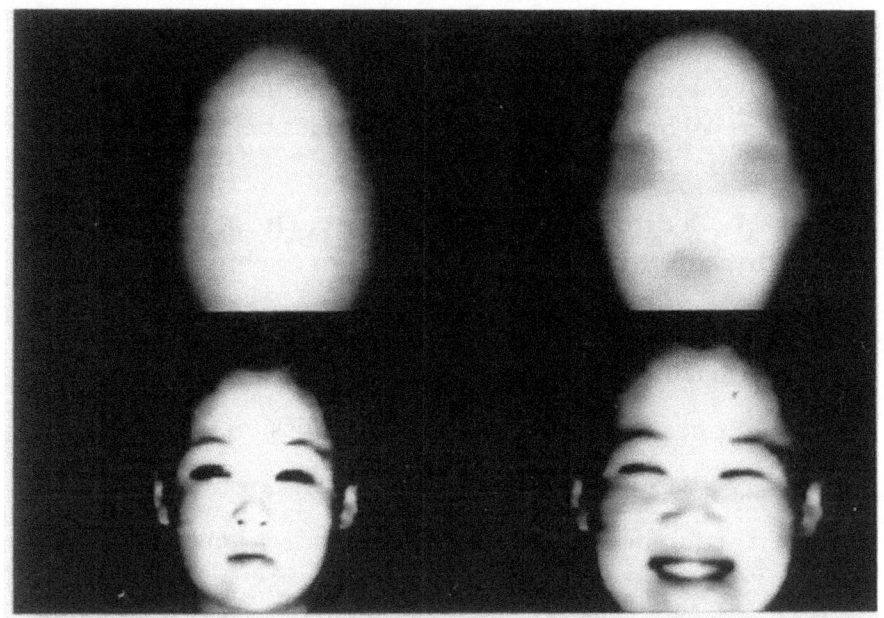

"Does a face really exist or a face may be an accumulation of infinite masks."

"Eventually it is the mirror coming from the deepest part of a human being."

"What is that? A message from heaven? Everyone can and must decode this message in its own way."

"These are all trials."

At Düsseldorf-Overath, in 1997, during a test with the Multi-Oscillation-TV-System, many images were sent from the invisible with the following words:

"We transmit by the crystal."

"We see the destiny of your Earth. We could share it with some of you."

"We are a transmission of energies. Energy-bodies can we be."

"We send you a power of Love. Accept it."

"Remember my face."

"We will always be with you, angels and guardians can we be. We are a group of energies being in resonance."

"We live in peace and expect peace. The one who disrupts the peace will have to bear the burden."

"We tell you, all those we are here, we help you."

"Amazing, we get to see you all, we know some."

"We need few words for the transmission."

"The Earth brings fear."

"Death, life—everything is reflected—a game

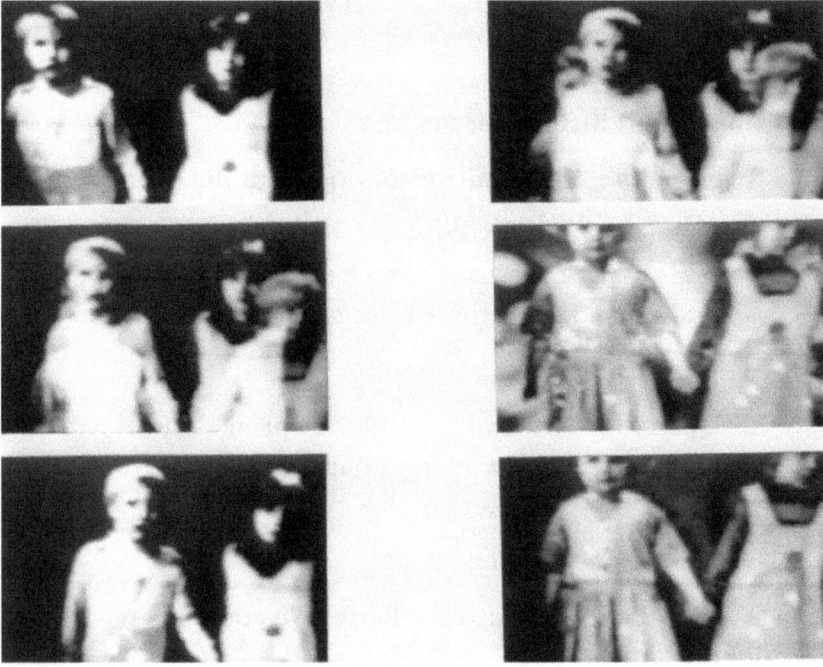

Children

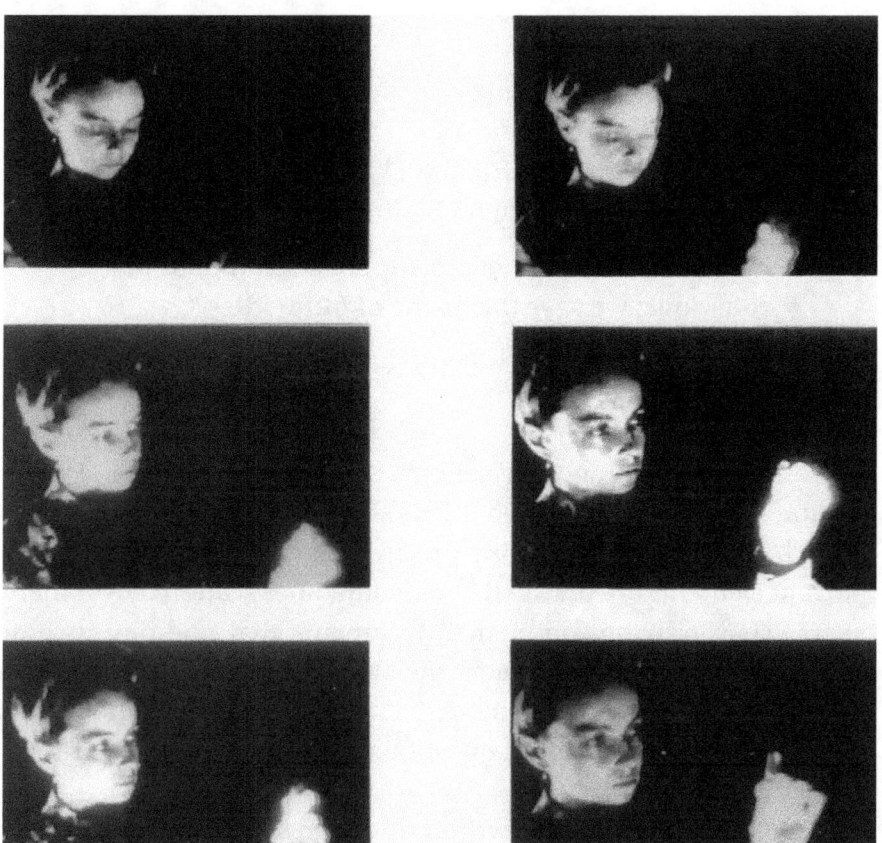

A woman with a crystal

How is the invisible able to show itself on a monitor by images such as those we have just shown? Here are some ideas:

> "We are thoughts; we show ourselves as you wish or want us."

> "Do not forget, we can be at the same time plant, animal, animal-human, like you or we want it. Every living being shows itself as you know it."

> "These images that we have saved in the superconscious and the unconscious."

> "Here we have the possibility of taking the form that corresponds to our representations."

Or the two answers received in 2003 through the HRS-System to questions:

> H.O.K.: "How come that you can show yourself during the experiment with the TV monitor as you were when you were living on Earth?"
>
> **"We save the image in our superconscious and unconscious; we save the image of our bodies."**
>
> H.O.K.: "Where do you pull the force to show you?
>
> **"All is energy."**

Understanding the phenomenon requires additional information. The researcher stated, "This strategy is typical for the Spirit World. You will never get any precise and extended response. Important elements are revealed to us, sometimes scattered over many years. They transmit only small fragments that one day give an overall image that illuminates the whole and makes everything clear. We realize every day more and more things we do not know. So internally we ask a lot of questions, while our minds constantly build hypotheses."

Generally, the researcher refrains from speaking openly about his hypotheses on these phenomena as long as he cannot verify them or present them as extremely probable. If not, "How should a layman get his bearings with such techniques someday?" he often says, with a shade of sadness in his voice. "This has already become such chaos that I prefer not to add to it."

In my interviews (the authoress) with H. O. König about his contacts with the "TV-Generator," it's striking that this stage of the journey didn't interest him any longer. Moreover, he has no desire to talk about it. "I was used to receiving a lot of criticism, but during my tests with the TV-Generator and its very good results in image and sound, the critics were so numerous that the work to get there was colossal—there was no comparison. With the advent of the computer and its many technical possibilities, I would have spent my life justifying my work against the attacks of fraud instead of using my energy in my laboratory work.

In the world of ITC, discussions became, from my point of view, often absurd. And suddenly many other so-called paranormal images appeared everywhere, but I was sure most of them were jokes. Carried away by the enthusiasm, we do not want or are not ready to look critically or establish the important distinctions between all the observed phenomena. So I no longer want to continue this line of research which, in itself, is still fabulous but requires a huge investment of time and resources. To be up to speaking of it with full knowledge of the facts, it would've been necessary to do a great deal more experimentation and research. Nevertheless, I am very grateful for all that I could discover by using this system. The images combined with sound are there. Nothing else matters."

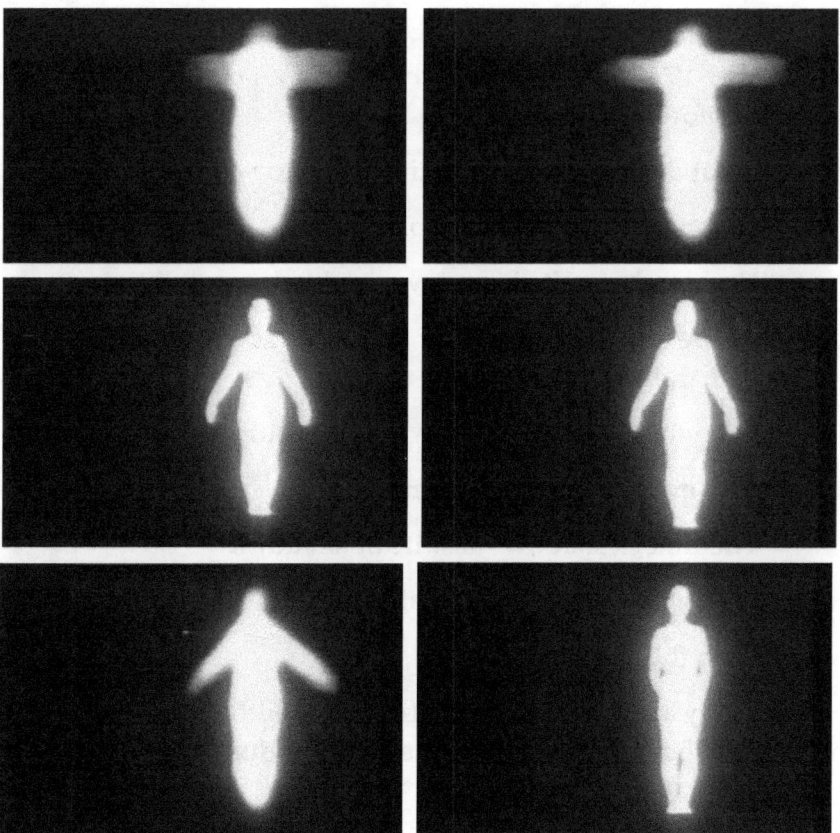

Paranormal photograph   Photograph, when he was alive

In December 1994, when we saw the face of an elderly person, we heard the following message:

> **"What does not fit in a lifetime? "**
>
> **"Hazard? No. I must say: no."**
>
> **"Now we know it all, and we need to know it."**
>
> **"By the hands of a superior direction, as long as we want and we do the proper thing."**
>
> **"Everything else collides in itself and of itself."**
>
> **"And everything was yesterday."**
>
> **"The time, a shadow—a suite of shadows."**
>
> **"Follow the shadows is finally all our lives."**

Let us listen to the following words:

> **"Many are sleeping on the first level. Others believe they continue to live as they did during their life on Earth."**

In addition to the last message the received image in movement

The HRS-System Based on Crystals
(The Hyperspace System)

For years, the researcher made daily trials with plants as well as with crystals. He experimented for a long time with philodendrons. Do they react to certain external influences? Could he use them to make contact with the Invisible? He slowly learned to discover some of their individual reactions. For example, one particular plant always reacted only when his son entered the laboratory, and another one when he himself set foot in his laboratory. Both plants loved classical music, but not the same composers. But among all these observations, the researcher did not find a useful way to integrate them into his fundamental research. He is sure, however, that plants, just like crystals, can provide invaluable assistance in the "contact bridge."

For two years, he studied the crystals, gathering important evidence about them. They will play a leading role in the new system. The Spirit World transmitted this work to him, and often drew his attention to their importance:

"Crystals are the key for the connection."

"The path by the crystals."

"We will come by crystals."

"Take the crystals as support for the contacting."

What kind of quartz should he work with? "**Rock Crystal**" was the reply. So he began to delve into the mystery of the rock crystals. He examined them intensively and discovered some of their secrets, both through technical experimentation and through meditation. Thus he penetrated a world as vivid as the world of humans, animals, or plants. Using technology he was able to observe objectively how these crystalline beings slowly revealed to him knowledge often forgotten today, but found in many cultures and ancient writings.

What extraordinary powers do they possess connecting the visible and the Invisible Worlds? The crystals became real friends of the researcher, who had beautiful personal experiences with them. But to be able to unearth their secret life by objective analysis, he had to invent another device. So he began constructing a system with transistors, photos, and diodes, allowing him to observe how and on what frequencies the crystals reacted. Several years passed, with daily tests to explore their features and lift the veil of their operating systems.

According to the researcher, the crystal is a living being that reacts to external influences. It can be stimulated purposely and has certain characteristics that one can even calculate mathematically. On what frequencies do crystals react? This was the question he explored more deeply. To discover the answer, he illuminated them, for example, with the whole color spectrum and infrared rays, then observed their reactions. One night he received a message from his spiritual friends directing him to use ultraviolet light.

He followed this advice and found that the crystals did not react to UV rays A and B, but to UVC rays of 280 nanometers."Due to the radiation," he explained, "the crystalline structure is activated and begins to change. In addition, this element has a certain point that must be activated and resonated in order to establish a potential force. By this activation, the crystal releases a certain amount of energy to support the contacts."

He noted that each crystal reacts differently on the same frequency and each has its own frequency, which can be analyzed and measured. But if these powerful elements have a life of their own, do they also act in relation to each other? Do any of them resonate with each other, as is the case with humans, some of which vibrate on the same frequency and others do not? The researcher studied, so to speak, their social behavior. His goal? Find six crystals aligned to the same frequency and harmonics to create a more powerful force. This idea came to him when quietly listening to music, when he suddenly remembered his musical knowledge and a message he recently received:

> H.O.K.: "What can I do to establish a longer dialogue?"

> **"The crystals in appearance with meaning."**

That inspired him to find a tuning of six frequencies, like a concert of harmonic oscillations. Therefore, he had to find six elements that vibrated at the same frequency: one crystal setting the basic frequency and five others that reacted harmonically on that basic frequency. Eventually, he found six among fifty different crystals, after examining, analyzing, and measuring them. He used this harmonious ensemble during his usual contacting procedure, placing it close to his radio while activating it by rays. The contacts improved immediately.

At that precise moment, he received an indication to go further in developing this system. He understood that in this apparatus, the crystals would provide the necessary support for the modulation of thought coming from the invisible entities. But he still had to find the most important crystal, the seventh one, able to capture the entire concert of the six others, gathering them and projecting them, as a kind of antenna, back to the Invisible Worlds. A new technique was born by which the crystals supplied the necessary support so important for connecting with the Beyond. In the years that followed, this device was changed and improved for each new public experiment. But all variations of the system demonstrated that the crystals were a fundamental aid in establishing contact with other life planes. **"We get into touch by crystals."**

This teaching of the spiritual world couldn't be more real and true, although completely new and impossible to conceive of for any mind that still abides in conventional knowledge and the laws of established physics.

The researcher's wife, Margaret König, dubbed the new system the "Hyper-Space System" because it allows us to transcend our world of space-time to connect with other existences far from our terrestrial universe. It was the first time other universes that had nothing in common with the Earth plane had the opportunity to make their presences felt.

Through the crystalline beings, entities came to meet us to establish a real dialogue with voices of incredible clarity. After so many years of exciting adventures and hard work, finally Hans Otto König had reached his goal: considerably improving the sound quality and the length of messages. But for many people who had not closely followed his progress with contacting, confronting these voices so similar to the human voice far surpassed their understanding. They could not believe that these voices came from other worlds. It was too difficult for them to comprehend this phenomenon; the leap into the unknown was too huge. The researcher realized the profound truth of the words transmitted by the Zentrale:

"Listen! We know your questions. Only a few do it understand."

"In fact," the researcher explains, "it is like trying to explain higher mathematics to a child of six who has learned only simple arithmetic until now—except that openness to this phenomenon does not depend on the intellect but on the maturity of the mind. It is, therefore, obvious that the child turns away from it."

But there are many people who already know higher mathematics, to prolong this topic, and now expect to develop experiments with greater systems. But during these trials, you never know in advance whether they will succeed or not, or what will happen. As the researcher points out each time:

> Never forget that these are all trials, ideas made for new trials whose results are never known in advance. We are faced with the totally new. Nobody can be sure of success. I know already many parameters that come into play, but there are still many more shadows than light areas. All these tests, however, allow us to learn a little bit more each time, even those that fail. Success is not essential. What we learn from experimentation is decisive.

All the many live experiments with the new system, however, succeeded in every conference or seminar. On the other hand, not all the transmitted statements are easy to acquire by human reasoning: one communicating entity is said to come from the star Sirius. So even if, for the most part, we are here switching completely to pure science fiction, it is important to take note of these objective, measurable acoustic documents that at least three hundred people heard live some afternoons in a small German town named Wesel. Do not blame us for disrespecting reality because it would disturb our worldview. Just keep your eyes and mind open, and remember, if necessary, Paracelsus's words: "What is considered by a generation as the pinnacle of human knowledge is often regarded as an absurdity by that which follows, and that which is considered a superstition for a century can form the basis of science for the next."

"Looking closely at the information transmitted by the Zentrale over the years, it absolutely does not seem so strange to conceive of communication with other levels of existence living on other planets, even if they are, for example, 4.6 million light-years away from ours and the information would take hundreds of years to reach us:

> **"Listen! The universal conception of time does not exist. One hundred years for you are like one second for us. There are other energy shifts that exceed your physical laws by trillions of years."**

"Listen! There are other energy shifts that exceed your physical laws by trillions of years."

"Listen! The cosmic communion is a possibility to have contact with extraterrestrial life."

"Listen!: On other planets, there are other life forms. They try to get in touch with you."

"Your attitude is different."

In view of all the messages received, it seems, on the contrary, more and more pretentious to limit ourselves and our universe to the borders of our five senses and to accept as real and true only what emerges from them. The adventure of the human mind seems much more vast and fascinating than what we are led to believe. But this would not be the first time in human history that man is offered the opportunity to leave his cave but prefers to remain inside. All changes at the level of thought are made very slowly on our globe, which is after all still so young in the face of the birth of the universe, and our culture is still undeveloped from another perspective. Future times perhaps or certainly will give more insight into all these issues. For now, let's observe and collect the data.

But is it true that we are surrounded by mysteries and unsolved riddles that should teach us modesty and humility before that unfathomable that is behind everything visible."

H.O.K: "What should I do to develop myself in the afterlife?

"Everything you learn here, you can use in the afterlife. Nothing is lost to you."

H.O.K: " Is it important to have so much pain at the time of death?"

"The hardest death is that you consider the easiest death, that means the unexpected death. It is possible that the soul of a defunct man has to get, what we call a shock of the soul." (Trauma)

The UDS-System (the System at Universal Direction)

We have arrived at the last stop of the journey of exploration: the invention of the UDS-System, or the Universal Dimension System, which the researcher is changing during the writing of this book.

We could say, basically, that this unique system brings together two others that preceded it: the Generator and HRS-System. For this technique, the Invisible World transmitted six new frequencies to him.

Several changes have already occurred along the way: the "crystal-mother" has been changed and the rock crystal have been replaced by six Herkimer diamonds, quartz-mono-crystals, the only crystals that are not born in colonies.

With this system, the researcher tries, for example, to better understand the influence of crystals to identify how the selection with the Invisible World is made and to create a broader frequency spectrum so that multiple entities can get in touch and communicate longer.

Experiments with this extremely sensitive system are done only in the presence of people who show a profound interest in the Spirit World and who are ready to prepare themselves for this exchange. By this system, it is no longer the spheres where the consciousnesses of the deceased are transmitted, but other supra-terrestrial spheres populated by various energy forms who have never lived on our planet Earth. We are grateful for what we can still discover, live through technical systems, of other dimensions that surround us and with which we are intimately connected, whether we are conscious of them or not.

"It has been about three years since I wrote (*the author*) these last lines. During this time of intense research, there were new insights, many individual sessions, and wonderful recordings and encounters, both through conventional methods and through the UDS and Infrared Systems. This facility was reactivated 20 years after the Spirit World transmitted to the researcher six new frequencies, which are connected both to the spiritual structures of Hans Otto König and myself *(the author)*. Thanks to this system, in the summer of 2016 we received, for example, the following wonderful statements:

> "We will try to close a contact field you to A.M. Wauters and H. König.
>
> Contact field closed to A.M. W. and H.K.
>
> Man on Earth now has reached a point where destruction would be complete without the help of higher spheres, both spiritually and physically, the entire humanity of this planet would be lost.
>
> Listen!
>
> What we perceive are your thoughts!
>
> Focus on your thoughts!
>
> We will register, record and save them!
>
> With every single person in this room, we will make contact in dreams, in your thoughts!
>
> Listen!
>
> We welcome the group here in the room.
>
> We'll try to contact each one individually, from mind to mind!
>
> Each one of you will get some information for your life!
>
> "Listen!
>
> We are seeing loved ones, being close to you, able to contact by your thoughts.
>
> The Earth now shall be placed in a new state of consciousness.

Its power of consciousness shall be raised.

Listen!

The life after your earthly life will mean that we'll change one of the various view windows again, which once was dark and opaque for you, then will become transparent.

Listen!

You are part of a vast consciousness contingent.

You can realize it by meditation.

Listen!

Everything you're doing will be observed and recorded.

Listen!

Consider that you are instruments of Divine Evolution, Spiritual Reality, an extension of Material Reality.

We are closing the contact.

Energy is coming to an end.

We send you an Energy Field and a Force Field."

Here again we listened to the statements of the Zentrale, being highly developed spirits, saying that they have gone through our whole evolution and therefore know well the realities of our planet. But, as announced, the UDS System created some intellectual structures or light beings that have never lived on our planet and whose "residential areas" are infinitely distant from the Earth. Sanaedes, coming from the fixed star Sirius and being familiar with us for a long time, no longer got in touch with us acoustically through the device.

Here in the terrestrial realm, however, her messages still cause great turmoil, again bringing the researcher a lot of hostility and reproach. Someone accidentally discovered that several sentence fragments expressed by Sanaedes during the recordings could be found in a book written by a very beautiful woman, Gerda Johst, who during her lifetime was in direct contact with the Spirit World. This book, entitled "The Uncut Jewel" (Reichl Publishing House) was later read by Gerda Johst herself and then published as an audio book.

The beautiful voice of Sanaedes has the same timbre, or "sound color," as the voice of Gerda Johst. Although Hans Otto König knows that similar phenomena have happened in the past with several experimenters, such literal transmissions were unknown to him in his personal research. How is it possible to distinguish a human voice from a paranormal voice when listening acoustically? Using a highly specialized process and apparatus, we can now measure individual spectral-frequency images and analyze and verify paranormal results.

We can therefore recognize and observe very well that acoustic messages differ measurably in spite of acoustic equality. And why do they only occur through a spiritual entity who is foreign to the human language? We take a little step forward in understanding the great mystery: how can a spiritual being who has never lived on the Earth articulate itself in an intelligible language that does not exist in its own world at all? How does she manage to transmit her information and realities to us? They use pathways that require a lot of time for us to understand them. To date there has been no effective explanation for such extraordinary phenomena. But, as already said, the whole of ITC research is a series of "scientific anomalies" that do not subordinate themselves to the laws of material existence and that, according to human reason, cannot and must not exist.

"But they do exist!"

And fortunately they do exist! They convey to us insightful messages, inspiring and challenging us to continue to reflect and research.

They show again and again that the research of mental structures will never be subordinated to the logic of our five senses and mechanical physics. It works with structures existing beyond the tangible and empirically ascertainable laws on the subject. But what laws do they work for? How often in his 48 years of activity did communication with other life planes leave the researcher completely subdued, amazed, and fascinated by what was happening in the recordings? How strongly did he realize how little we know here on Earth about spiritual realities, about Invisible Worlds and their interactions with our earthly world?

In his search for answers, this scientific mind has often encountered the limits of his own reason, rationality, and imagination. Since his encounter with the paranormal voices in 1974, he has often been asked to accept the unexplainable as a probability and to accept unimaginable events without being able to explain them immediately. But in those 48 years he has received many answers to many questions. Countless small technical discoveries shed light on still-unexplored new land. He had, however, slowly and patiently developed these new insights himself through titanic work and a very intimate connection with the Spirit World. Fundamental research moves into an incomprehensibly strong field of mind, a fascinating adventure that shows us again and again how small is human consciousness and how great and strong is Spirit.

Human beings are always inspired to open up to new realities and to wait patiently until we get more knowledge about these true messages. Perhaps one day serious researchers will be able explain to us the explanatory models of these extraordinary phenomena that have occurred over many years. Let us, however, continue to listen with great attention to the laity about what other realms will convey to us, and be very grateful that spiritual beings are able to contact us acoustically. Who will tell us about the great apparatuses, and what will the Spirit World transmit to us? From which life planes will we still receive messages? How many people will the Spirit World accompany lovingly during their spiritual growth, sustaining them through their whispering voices and their acoustically, directly perceptible presence in the recordings? Let us remain open, not hesitating to question, and try to get answers to our questions: Why, wherefore, and for what reason?

The answers, however, are never given to us immediately, like a ready-made recipe, but belong to the future.

Yet a huge breakthrough has already happened: a collective wall has been pierced by the power of one human mind in connection with the Spirit World, the dark has been illuminated, and a spiritual-technical contact bridge has been built, where many spiritual entities can directly manifest through various technical systems in a very clear manner. This fact, tremendous and highly fascinating, is still rarely realized on Earth. The fire of knowledge burns around us, true reality irradiating our dusty plane with light and hope. Hans Otto König's research is a powerful, valuable, and unique pioneering work, which, like any pioneering work, is made for the future, already preparing itself. It will always give us new acoustic pearls from other worlds directly connected with ours. Finally, let us hear the words of those who have accompanied the research of Hans Otto König since the very beginning: The Zentrale. In July 2016 they transmitted themselves again, very clearly audible through the Infrared System. The circle of a spiral, lovingly closed ... till the next ....

"Listen!

The connection to us are windows to new insights and let you intuit what you can experience when you open yourselves!"

"Listen!

Now you have a real proof of our reality!"

Third Part

The Messages from the Spirit World

""Let parents bequeath to their children not riches, but the spirit of reverence."

~ Plato

Introduction

Since the beginning, man has tried to understand the universe, penetrate the secret of existence, and find answers to the big metaphysical questions. To achieve this, he followed many paths: those of science, religion, mysticism, philosophy, trance, and many more. It is said that each time the Spiritual World came into contact with some open-minded humans, able to hear, it transmitted to humans knowledge not limited to the Sensible World. This knowledge, collected in the privacy of inner silence, ended up dispersed almost everywhere, conveyed secretly, hidden in multiple and diverse writings. It is present in all civilizations, revealed in all major religions, expressed in the works of the mystics, alchemists, and ancient philosophers, and written in myths and tales. This knowledge arouses the curiosity of each pilgrim.

During the twenty-first century, a century in which technology dominates, Parallel Worlds are unlocking new knowledge through technical devices. They help man build a **contact bridge,** where many spiritual entities can whisper their words directly, bypassing the body of another human speaker. This path is new and very different from that of mediumship, Spiritism, or channeling, where people often doubt the reality of the contacts, and where the spirit of man can interfere or distort the message according to its subjective worldview.

The technique makes it possible for us to directly hear the Spiritual World, to receive their messages without any deformation, and to examine these voices for their authenticity. The differences are daunting. The ITC offers man the opportunity to move towards wisdom and knowledge from a real and concrete foundation and without prejudice. Technical devices play a very important role, but they are also very difficult to develop. Understanding this acoustic phenomenon unfortunately progresses slowly, and the energy available to transmit information still remains limited during the contact. H. O. König's longest contact was about 20 minutes. But hearing the Spiritual World address humans directly, and feeling their presence in these contacts, is an extremely valuable and beautiful experience that life offers to humans, often lost on the Earth plane and questioning the meaning of existence.

Who is calling?
The Invisible World is Full of Diverse and Multiple Spirit Beings

Who is sending the messages? An inevitable question!

"It is the Invisible World of Spirit," the researcher affirms, "with its multiple information fields. In some cases, through historical investigation, I could verify the identity of the sender. They were always people unknown to me, but who had actually lived on the Earth plane. In other cases, they were people I had known personally. Or they were people who knew the person who began making contact with me.

But I also got a lot of names and messages that I could not attribute. When I get a message like "Here Antonius, I am Antonius," what can I do? The development of my research, my technical systems, as well as the content of the statements, are the decisive factors for me. They allow me to advance in understanding of the phenomenon. The image of the Spiritual World becomes clearer and is clarified gradually through the numerous contacts. I can always recognize a "psychic structure" by the wording of the messages given to me. In addition, the backup of the spectral images allows me to observe whether it is the same "psychic structure" talking to me. I observe, note, and reflect on the received data.

Communication with Loved Ones

Many messengers often give the name they had during their lifetime on the Earth plane, very often first and surname. They assure the researcher that they are still alive despite their physical death...

"Here speaks the deceased Gunther."

"Carla calls."

"Moris from Gladbach."

"Here calls Paul Seeger."

"Here Anja, I am still alive, happy here."

"Leni Shade greets, lives here with many friends, greets Gretel."

"Here Hubert."

"Birgit calls."

"Hello, here Doris lives."

"Doreen Bauer together with her brother. Even the dead now call."

Communication with the Animal World

Among the communicating beings, there are those who say they have lived on Earth in the form of an animal. There are many experimenters receiving signs of life by animals passed into spirit. H. O. König himself has a very deep bond with animals, which he considers his equals. He is convinced that animals have consciousness adequate to their degree of intelligence. He established with his Yorkshire-Terrier Julchen a kind of telepathic communication, and the little dog came to perceive the invisible entities better than most humans. His male cat Stanislaus always knew when his two-legged friend was coming home, even though he returned at different times, and waited for him at the front door. His animals manifested themselves to him repeatedly after their death by giving their name: "Julchen here, everything beautiful here."

To the researcher's question "What is the third level?" came the reply:

"Where humans and animals live peacefully together."

Communication with the Extraterrestrials

While most of the "information-thoughts" come from entities who had an earthly existence, some entities never lived on
Earth. Long before it would ever happen, the Zentrale announced
to the researcher that one day other energy-forms would try to get
in touch with him:

> **"There is life on other planets in other solar
> systems. They try to get in touch with you."**

Because of the integration of crystals in the HRS-System, dialogues developed between the experimenter and an entity called
Sanaedes during many experiments in front of a large audience. In
a beautiful, clear, feminine voice, she said she came from the star
Sirius:

H.O.K: "What is your name?"

"Sanaedes."

In two separate tests at different times the experimenter asked
where she came from:

> Sanaedes: **"I am coming from the Realm of the
> Stars. From the Fixed Star Sirius."**

H.O.K.: "What is the importance of the Star Sirius?"

> Sanaedes: **"That the Fixed Star Sirius is a star of the
> highest importance, where the most elevated minds
> come together for discussing earthly questions and
> problems."**

H.O.K.: "Where do You live?"

> Sanaedes: **"A wonderful supernatural world, where
> we are living. But it will be too difficult for you, to
> understand."**

H.O.K.: "Can you describe your world?"

Sanaedes: **"My world is the World of Dreams and Beauty."**

H.O.K.: "The last time, Sanaedes was introduced to us. What is your function?"

Sanaedes: **"Sanaedes, Ändyminijang, Anecebio, Mandenmides, Millimieres, these are names of angels, who are there only for the beings of the Beyond. Therefore they do not make part of the Destiny Angels."**

She says she would never have lived on Earth, seeing our planet from a very different perspective. Her reality, however, is beyond our understanding.

Can we get in touch with any entity living in invisible universes?

"Which entity makes contact," the researcher explains, "depends to a large degree on the vibration of the two communicators and/or the proposed vibrational base. The whole system of Instrumental TransCommunication is founded on the principle of resonance. It is a circular connection between the experimenter, the technique, and the entity of the Spirit World (the transmitter plus the receiver)."

The possibility of transmission is also related to the technique used. Because of the proposed vibrational basis, certain entities can manage to communicate with us while others cannot come through. Any communication system does not allow all entities to manifest themselves acoustically in our world.

Then there is the "Time" factor that comes into play. Getting in touch with a person who has just died is not at all the same thing as contacting a person who has been dead for twenty-five years, for example. And among a great many people who have just left the earthly life, all do not communicate with their loved ones. Reactions can be as varied as the people themselves. Some Spirit beings are very eager to tell their relatives that death is not the end of life:

"Here a new life comes."

"Mom I see you, listen!"

"We live in love."

"Death-Radioing, we listen. But we are alive."

Others, on the contrary, do not manifest anymore. It seems to depend on the intimate bond people had during their lifetimes here below.

Some, having passed on quite some time ago, still feel the need to transmit something, as in the case of Gerd Siebenreicher, discussed in the first part of this book. Others completely turn away from Earth, where nobody is thinking of them anymore. They trace their own path in the afterlife without perpetuating earthly contact. Or they are just reincarnated, and the link is severed. In fact, everything seems to depend on the individual choice of every soul, on the task they take on in the other life plane, and eventually on the attitude of those still living on Earth. In so doing, they often try to bring themselves into mind:

"Do not forget us!"

"Here are many people, wishing to connect with you.

Many children are here, who greet their parents.

They are happy that you have come.

Don't be sad; not all are able to talk with you.

But you will get in touch with them, by your relation."

"Is it possible, that highly evolved spiritual entities are talking with us through the technique?" many people ask.

In Germany, many people are convinced that communication through technical devices would only allow the manifestation of haunting ghosts or earthbound souls.

There is an interesting excerpt from a French book titled *Les Mort Nous Parlent* (*The Dead Talk to Us*). The father, François Brune, talks about Instrumental TransCommunication by citing a passage from a spiritual teaching that a mother received from her deceased son, Roland de Jouvanel, through automatic writing: "Roland de Jouvenel had announced that one day this kind of communication would become possible. But he had also reminded us of the limits, which could not be attained by procedures already used, but by reaching the levels of the beyond, whatever means is used.

"Occultism and Metaphysics will become an experimental science based on 'The Real' (= the Philosophy). A raising table is a phenomenon of waves/interference; mediumistic conversations are contacts with Spirit beings, still close to the Earth plane. There is the phenomenon of interpenetration of one plane into another one, but this area is immeasurably far from our realm. These incursions of one plane to another will become as familiar to us as aviation nowadays has become common. . . . We will get to communicate with the Invisible, but these Invisible Worlds are as far from Divinity, as much as we ourselves are far from a star. . . . A day will come when we will capture the vibrations of the levels as easily as we have captured the power of electricity, and they will be perceptible for us. But God is still not there . . . The mystical or spiritual experience is something quite different . . ."

What does the German researcher think about all these statements, or would it be better to say suppositions? "It seems to me that this way of thinking about minds being tied to the Earth—and so at a level of inferior evolution—is a remnant of religious beliefs, or an idea conveyed in some writings, unless it is a way to stand at a distance from ITC by saying that, in any case, we can have any contacts with less evolved Spirit beings.

To my mind, however, nothing is more wrong. Knowing that the messengers do not stop telling us that all is one, it is very understandable that a highly evolved spirit nevertheless has a certain connection with the Earth plane and is able to transform itself to higher planes of lower vibration to transmit important messages to humankind. Christians would speak of this as an act of love. As for the mystical and spiritual experience, I remember well what it is. I do not need any spiritual, mystical, or Masonic pretension, or any alchemy, and I do not care to know if I can have contact with evolved planes within the divine power or not. That is not my way of thinking. Naturally, I asked them how they consider the question of God. That interested me enormously."

"My research is based on real facts and concrete results. It is what it is, and it is not for me, the researcher, to judge the information that I receive or filter it according to my personal convictions. I observe a phenomenon and try to understand the whys and wherefores. This strongly distinguishes me from all those who study the messages received from the Beyond. This research is still in its infancy, and needs care not to draw conclusions too quickly on certain subjects about which we know so little."

The technique has revealed that souls who have just left their physical bodies keep, for some time, the thoughts that were essential to them during their earthly lives. "These psychic structures" do not change because they have abandoned the physical body. On the contrary. The passage of abandoning the physical body is a unique experience unlike any other, but not does not change the existing structure of an individual consciousness. This will change gradually, as it encounters new experiences and realities in the more subtle evolution planes.

Although the researcher personally had very little contact with earthbound spirits, remaining chained to the Earth because of their exaggerated materialism, he knows that they do exist and that they were heard by several people during their radio recordings. As the life and thoughts of these spirits are centered around the acquisition and preservation of material possessions, they remain prisoners of the material, even after leaving the physical body. The Zentrale forwarded the following:

"Listen: Earthbound Spirit beings, as you say, are men, who are very attached to material goods."

People caring only for tangible realities in their lives as the means to enjoy their existence will be in store for big surprises once they pass to the other side. As the Zentrale ensures: **"Humans, who ... have nothing but enjoying their life on their mind, their astral bodies remain so weak and have so little power that they would be diluted by the cosmic rays."** Other Spirit beings may have a view of earthly problems without being chained to the Earth. They often turn to help one particular person, to give him energy and inspiration. **"Mom I am with you and help."** At the same time, they continue to evolve on their new life plan, devoting themselves to their new tasks. Some of them may even be at a very high level of development, as they explained us. "To my mind," says the researcher, "spiritual entities at a higher level of evolution are indeed able to turn down to a lower level of vibration. The reverse, however, is never possible."

The superior entities no longer have a personal name. If they appear by name, it's only so human beings, who need a name for everything, can recognize them during their meetings. The first person singular is no longer used at all: they say "We." The everyday problems of an individual particularly do not interest those entities any longer. They are too far apart from that. The content of their messages is always of another order, more philosophical, and addressed to the human race in general. It is a gift that the Invisible World offers to the audience during contact, via an experimenter they choose themselves.

> **"Listen: A good contact with us, with the Zentrale, get only that one, who will get in touch with us, not by personal egoism. It is a gift we offer to people who is in a holistic relation with all life."**

> **"Listen: Not everyone can make good contacts with us by special, more advanced devices. Stature and spiritual maturity are the premises."**

> **"Listen: Only that human can have good contacts with us who bothers to understand the universal laws."**

> **"Knowledge is a force that only few people will acquire."**

> **It is always the Invisible World who decide to open and close the contact window:**

"We make contact."

"Contact, contact field closed to Hans König."

"We close the contact."

"No more energy, end of contact; get in touch again later."

Moreover, only through the continuing development of the technique—the Multi-Oscillation System, the Infrared System, and especially the inclusion of the crystals—has it become possible to reach these highly evolved planes of consciousness so distant from Earth, and have them manifest to the researcher in longer and longer dialogues. These entities perform an important central function on the highest cosmic level, the seventh level, which contains all frequencies of the other levels, as they have explained to us. They have no name, so H. O. König and his wife Margaret named them, according to their function, **"Zentrale."**

Their messages to the Earth are usually opened by this beautiful word:

"Listen."

What are the function and significance of the Zentrale?

"Yes, there are human souls, developed to the highest point, which act like angels on Earth. They come from the spheres of light, and they empower you."

A question from the audience: "What is the significance of the Zentrale?"

"They are as cosmic forces not unknown to human beings. We could call them Deities; we could also name them Spirits, Forces, Cosmic Entities, which are accomplishing a very high function and which are inexhaustible power sources for human beings."

H. O. K: "And what is for you the Zentrale?

"Incredible, powerful, spiritual Zentrale."

"Listen: Our form of existence is the highest plane -
the Cosmic Level of All- Embracing Love. Anyone
connected with us in love will not suffer damage to
body or soul."

They say they have lived in our area on the Earth and so know the
concerns and thoughts of human beings:

"Listen: We know all your thoughts."

"Listen: All your thoughts are known to us."

"Listen: We want to tell the humans, we know all
our thoughts. All your thoughts are recorded."

The statements of the Zentrale touch our hearts as well as our
minds. They say:

"All is easy."

They never made long dissertations, but clear and exact statements.
There are no forbidden questions or contradictory messages.

"There is nothing hidden that will not manifest,
and nothing secret that one will not know."

These messengers of the Beyond testify to their concern for the evolution of our planet and the spiritual growth of human beings.
Here is the problem:

"We will communicate you all the concepts and
meanings according to your mindset."

"You get to understand so little."

Provisional conclusion: Who is calling?

We receive messages with names of individuals who actually lived on Earth, analyzing their spectral image with specific frequencies, all different from each other, to define a precise identity.
And the many words and phrases continuously affirm that this

"something" or the source of these messages comes from deceased persons now living in another world, parallel to ours. The dead tell us that they can remember all the details of their life on Earth. They read our minds, know our names, and respond according to our specific questions. Often their responses anticipate our questions. And very often the same communication can last years.

Then there are Spirit beings who reveal that they come from the Seventh Level or from the Fifth Level of Evolution. Others say they come **"from our future,"** or from Stars such as Sirius, Alpha Centauri or Vega.

To designate who is calling us from another world, we can use terms like "spiritual being," "entity," and "angel," or one of the more recent terms like "energy form," "psychic structure," "information field," or simply "soul." But are these terms appropriate? And what is the reality behind these words? Do we really know?

These questions, however, pertain to the material world. Do we know who the human being is? The researcher said, "If we can see that person, knowing his name, his age, his way of thinking, and if we have a long association with him, can we, therefore, claim to know who is talking to us, really? And if someone calls you, for example, on the phone, and tells you his name, can you really be sure that it's that person who is speaking at the other end of the line?"

The answers remain open for anything falling within the visible range as to who or what is living in parallel worlds.

The perception—or not—by our five senses doesn't fundamentally change the problem. People need evidence to find their way in a reality that eludes them. The essential point, though, always remains unknowable. "That's why," the researcher says, "we refer first to what the messengers send us. We take the content of their messages seriously and consider them a starting point for any discussion and research about them. There are thousands of very clear messages, witnessing that there is life after death. We should not deny them, either because they do not fit in our familiar framework or for fear of the mockery of others."

But, of course, it is important to carefully examine everything, not only what comes from the visible world, but also what comes from the Invisible World. It would be a mistake to take the messages for granted simply because they arrive from the subtle spheres. Deceitful spirits also exist. The Zentrale stated:

"But reflect that during the contacts there are very different mentalities. The verifying is important."

"Listen: Be critical in your contacts. Examine your contacts."

And the researcher added, "Recognition of the essential form, whether embodied or not, ultimately depends on our own mind. Here we are subject to learning or long recognition process, which corresponds to our own spiritual development. Then each person, depending on his upbringing, his education, his convictions, will accept certain messages and reject others, pose some questions and refuse others."

So everyone comes in contact with the Spirit entities he deserves, according to criteria—or I should say laws—only known by the Invisible World.

"Only the one, we call."

"We make the contact."

To believe that we will receive in-depth teachings or universal knowledge from all Spirit beings is a great lure, but it shows great naïveté. Nevertheless, it's undeniable that there are messengers with more advanced knowledge than others, as in the case of the Zentrale:

"Listen: Our life form is of the highest level, the cosmic level of all-embracing love. It is called Zentrale."

"Yet," the researcher says, "it is always important to keep our own critical thinking and never become dependent on what they tell us. On the other side, it would be pretentious not to dwell on the teaching of these Spirit beings."

After a pause for reflection he continues, "Often people expect me to answer existential questions because I communicate with the Invisible World. Circumstances are different. The afterlife is full of information fields or, if you prefer, full of very different minds, each having its own vision. Furthermore, my approach is much more open and much more modest. I refuse to have preconceived ideas about the answer I would like to receive. Whatever comes through is interesting and significant for getting an idea of parallel worlds, whether it pleases me or not. I myself still have so many questions for which I have no answer! Fortunately! But the fact is that this research enables us to leave all these old patterns of thought and knowledge and move gradually toward a new knowledge with a real foundation."

Does a universal truth exist, or only subjective truth?

"Considering everything I have discovered in my research," H. O. König explains, "I am convinced that universal truth and knowledge exist. The "Self" totally vanishes, making way for a union with the Universe, with the Divine, where any polarity is abolished, as the Zentrale described. But the road to achieve this is very long. First, the "Self," which still remains after physical death, must discover its own path and its own truth. In these other life planes, the "Self" goes through multiple deaths and rebirths, but not in any sense that we can understand. The universal truth is beyond our human comprehension. Humans are unable to understand it. Remain humble and in all modesty begin on the path to self-knowledge and spiritual growth. True knowledge will be provided to us when the time comes.

I would like to quote here the Zentrale:

"Listen: Many humans have great questions, and don't understand the simple."

And even,

"We have just said so many things; few have been understood."

How could the highest cosmic levels convey their extensive knowledge and the universal truth to us when humans have only just learned to read?"

"Listen: In us is eternal knowledge."

What do they communicate?

During a conference at Wesel (Germany) in 2006, someone in the audience asked, "What does contact with the beyond teach us?"

> **"It is for sure that we are surrounded by mysteries and unreleased puzzles, which should teach us modesty and humility for that mystery, lying behind all the visible."**

A brief introduction is needed, referring to this core idea of communication between two different worlds, one within the spatiotemporal zone, and the other beyond space and time. How are entities imperceptible to our senses and living in a world with unknown realities able to make themselves understood and convey their reality?

For the researcher, one question remains. The messengers have told him repeatedly that many spirits know the reality of life on Earth and that they can read our minds: "To describe the world beyond our senses, they purposely use words and images that are comprehensible to our human consciousness, which is not always easy for them: **"It is difficult to talk about the contact bridge."** They are capable, however, because they themselves once lived on the Earth plane."

On many occasions they used the words, **"Like you are saying it,"** and **"as you are calling it!"** But there are also spirits who have never lived on Earth and who talk of a sphere where the ways of thinking and living are remote from the Earth and beyond our understanding. Even if many facts seem too difficult to understand, as long as we are held captive in a physical body, it is obvious that supernatural worlds want to get into contact with human beings. They are the ones who meet and talk to us from another reality and invite us to another way of thinking that we are collectively still far from achieving.

In the end, for Hans Otto König the most important messages he has received are, "our personal consciousness continues to live in other worlds parallel to ours; that we are all integrated into a structure, a system, where ALL IS ONE; that we can find our peace and live a life in a community, in a system of BEING."

For the survival of the Earth, it is urgent that we ponder our way of thinking and actions, releasing our materialistic and selfish thoughts, and moving to a new, constructive attitude and approach to life and death, finding new paradigms for nature, animals and plants, and especially for our fellow human beings.

"Let us act according to this new vision and leitmotif based on the laws of the Universe and on all areas of life."

1. "Listen: Tell all humans that we are alive!"

"How clear and distinct is the information from the Beyond, how realistic the contacts with supernatural beings." Various messengers have expressed this all over the world: Physical death does not mean the end of life!

"Listen: death does not exist; everything is designed for eternity."

They try to give concrete, legible signs in the storm of life and its daily noise. On many occasions, they sent these thoughts to the researcher:

"I have proof, contact by the dead, born again."

"I am living here."

"I know the plane where I have hungered."

"We are freed."

"Silke says, Doreen Bauer is together with her brother. They are happy."

"Death is a new life; what a wonderful world we have."

"Albert greets his wife, Ursula."

"I send dreams by Margaret."

"Trautmann talking here."

"Listen: We are alive and are happy, when you will see us again!"

One of the first small dialogues exchanged through the Multi-Oscillation System is particularly eloquent on this subject. It contains in itself several significant indications that constantly return to all those who set out to listen to them in all the associations (in France, Italy, Spain, Portugal, Brazil, USA, Russia, India, and more).

One autumn afternoon in 1983, in a large conference room in Fulda, a city in northern Germany, a mother encountered her daughter who had been killed in a motorcycle accident.

During the lunch break at the conference, Hans Otto König had started his devices to conduct a simple stabilization test. A few people remained in the room, including a woman who had just lost her only daughter, as he learned later.

Suddenly and completely unexpectedly, without anyone asking anything, the following words came through:

"Hallo Mama."

Hans König had the presence of mind to react and asked:

"Who are you?"

"Anja Dohrmann."

The exact name was given, then a reference to the current situation. The Spirit being stated the full name of the experimenter, who up to that time knew neither the mother, Marlene Dohrmannnor, nor her daughter, Anja Dohrmann.

"I can see you."

"Hans König is my friend. By him I can get in touch."

"Tomorrow is my birthday, but not here."

Hearing these words, the mother was visibly shocked, and when her deceased daughter transmitted her date of birth, the poor woman lost control and began to sob.

"You must not cry," was the immediate reaction from the beyond. Then came this message with startling clarity:

"I'm still alive."

Usually, it takes some time before those three little words, **"I'm still alive,"** are accepted in their simplicity by the whole person:

"Listen: You are receiving all by your heart and transpose it into thoughts."

So we hear with the heart that which is difficult to understand and accept by human reasoning. The Zentrale states:

"Listen: Join hands with the cosmic force."

"It cannot be understood by your consciousness."

"Only your heart can understand."

This was the attitude of Dina Tölke, a bereaved mother who met H.O and Margaret König after the death of her beloved son Franck. She often withdrew in silence to communicate with the Invisible World. In the following passage, she is contacting her deceased husband, Heinz, who, during his lifetime, could never accept the contacts his wife established with their deceased son.

Dina T.: "Heinz, have you met our Franck?"

Heinz T.: "Our Franck is alive."

Dina T.: "How are you doing in your new life?"

HeinzD.: "My death is my great experience."

Dina T.: "Heinz, what do you think now of my contacts?"

Heinz T.: "Now I understand it."

It takes a lot of enthusiasm and patience to accept that invisible waves passing through our walls and homes are capable of influencing our technology and can bring us marvelous gifts from the World of Light.

In this final example, contact was made in 1989 between the Invisible World, H. O. König, and Marlene Dohrmann.

"Contact."

H. O. König: "Hello, friends!"

"Contact for Mr. Dohrmann and H. König."

Then another voice, with the same spectral image of Anja Dohrmann, continues the dialogue by saying,

"Mama, Hans, are you able to hear me?"

H. O. König and Marlene Dohrmann: "Yes, wonderful."

**"Me too. We are all here. Everything is in order
with the technique. We are waiting for you."**

Then the first voice again, concluding the little dialogue:

"Contact end."

Reason often loses ground in the face of real experience, apprehensive of being subjected to an illusion. So it draws itself up proudly, defending itself against that intrusion, disturbing the course of its familiar universe and its acquired knowledge from a solid education. Most people are inclined to give more credence to human scientific knowledge than the subtlety of their own experience. But the Spirit World is concerned very little with the authority of scholars.

"Listen: Quarks, atoms and quanta are of little importance to us. Only your maturation of mind is essential. The mind has a different meaning. Everything is simple."

Seen from the world of stars, humans have often lost the ability to think for themselves. Direct, unique, and personal experience with reality has become a rare thing. Instead of seeking the truth, we often follow or worship terrestrial idols of all kinds, whether through fear or habit, or familial or social conditioning. The Zentrale insists:

"Listen: you must learn to think and to act independently again."

But you meet few people who still resemble the child of Andersen's tale, daring to cry out: "The emperor has no clothes." The battle on Earth is often difficult. It is not easy to find a way to get a place on the big stage of life. So social issues take precedence over the innocence of the gaze, the noisy speech over fine listening, and fear over the courage of the authentic thought.

Through the centuries Spirit beings have continued to visit the lower vibrational levels of the Earth plane, to meet with us and promote our spiritual growth and strengthen our knowledge:

"Listen: No life will end!"

"Every life begins with new experiences."

"Prepare yourself for the life to come."

"Learning the laws of the universe are premises."

Human beings willing to listen are a minority. Caring for the realities of the spirit seems to be a luxury in our modern age.

"I'm always calling; no one hears me."

Modern people have forgotten the song of the Spirit, capable of delivering us from our chains and physical finiteness, and carry high the banner of the material and carnal lust.

The Zentrale reminds us:

"Sad when nobody listens."

"We have said everything."

"The body will be transformed; the soul, the mind exist eternally.""

Usually, we focus only on the realities of the afterlife in times of bereavement or illness. Then questions suddenly arise, like "What is the meaning of death?"

"Nothing is dead. Do you cast off a worn-out coat with tears? The death that you regret does not exist."

Going toes up is a reality often difficult to accept; coping with the loss of a loved one can be extremely painful. Unbearable silence. A terrible void. An indescribable separation. No one escapes the unspeakable grief of losing life that we have spent in loving harmony with someone else. A warm, loving presence, then suddenly everything is black, cold, silent, vanished, only a nothing. Nothing more than memories in our hearts, keepsakes of the loved one now passed into another life plane. What actually are we weeping about? The life that was lost? The loss of our life together? Ourselves? The connection? The emotional life with the other one? The togetherness?

Grieving for a loved one is essential, important, and necessary. It takes time to find a new living situation when the beloved person is no longer physically present. The pain of loss is there, present and inevitable, but we can gradually lose this if we know that the other one is happy and doing very well.

The deceased, too, must grow accustomed to all the new circumstances he meets in his new life. Initially, he remains connected in thought with his loved ones, sensing their grief, their thoughts, and the vibrations of their souls. He knows that the human need for comfort on Earth is great. So he often desires intimate contact with the loved one, caressing him lovingly and gently by the hand, comforting him, sometimes even relieving a heavy burden of guilt, speaking things never said, or sending the power of love:

"Silke is alive; it helps the tears, you know."

Listen to what these children say, whose parents were present in the hall during the experiment:

"Listen: Some are here and want to sustain their
relatives: Anja, Stefan, Frank Werner, Birgit, Andre,
Andi, Silke Thomas, Helmut, Leni, and Otto Josef.
Here are standing still many others."

Highly evolved Spirit beings, however, do have the following messages for humankind:

"Listen: Many are crying for them, not for life that
is in trouble. We must tell you. You know your
loved ones are in good hands by love. Grief means
a new suffering for humans who have started a
new life."

"Listen: You know, they have fulfilled their task in
life. They have new tasks now, and you too."

An invitation to think about and to live death in a different
way. In which part of our ego are we imprisoned by our tears and
suffering? The words of the Zentrale push it towards a transcend-
ence of the self and its ego.

But our deceased loved ones will never judge that self, and
they will continue to send us strength and love. Lovingly and patiently, they await our arrival in the other world. And when the
time comes, often terrifying for humans, the Spirit beings stand by
the one who must die. Spirit bodies, usually too transcendent to be
seen with our eyes, often become a reality in the last days on Earth
for the one who will die. Those who welcome the deceased on the
other side are already there, with him and close to him, only a
breath away.

Their attentive presence allows us to cross the threshold of
death without fear and carry our soul towards new horizons,
flooded with light and brightness.

"Listen: You would feel a breath of Eternity.

To understand and accept the transformation of the
shapes of your minds, so you need not be afraid
any longer.

Fear and ignorance dominate your life. Only a few among you understand.

Understanding means living."

2. "Listen: All is One. Understand!"

"You have the opportunity to grow spiritually and recognize that in life, all is one."

This truth of "All is One," already revealed by the ancient Greeks, is the basic alignment of the great symphony of life and the opening overture of the silent connection of all living things within different vibrational worlds, between the Earth and the most distant galaxies.

"Recognize that we're in affinity.

We are connected with you in love.

Form a bond with the Universe.

All is one. Understand!"

The many messages let us become aware of the union between earthly and heavenly realms, and the importance of that connection.

It bears repeating: It is the Spirit World that is wanting and establishing the contact.

"Listen: We repeat. We need your contacts and your love, as you need ours. Be aware that we form a unified whole."

"We thank you that you answer again and again."

"We thank you for prayer."

"Listen to the Beyond, please!"

And the Zentrale affirming:

"Thoughts are the language of the soul."

But how do they match it to perceive us?

**"Every vibration of the soul translates itself
instantly into words."**

"I hear your conscience."

H.O.K.: "What is the energy carrier that allows me to
get in touch with you?

 "They are cosmic rays."

The Spirit World is not cut off from the Earth plane.

**"We hear you and see you. There is a field of
thought from you to us."**

The inhabitants of the subtle spheres are capturing that
blooming of our inner garden and reading our thoughts. Everything is mind, but on diverse levels and in different bodies. Invisible stairs are permitting man to ascend to higher planes, where
Spirit beings are living and descending to the lower strata of denser
matter. This bond of love, emphasized by the Spirit World, does not
interfere with those having to discharge other duties according to
their level of development.

It is a natural connection between worlds of **different material qualities,** united by the power of love:

"Love makes the wave and the transmission."

According to the messengers, we are just starting out on
communication with the invisible. The final goal of the contact
bridge is still unforeseeable:

**"Listen: The possibility of the contact bridge will
grow, and we expect an evolution. We ask for
commitment and help.**

Greater ways will be prepared in the contacts with us."

In every human being, there is a constantly alert and activating part that never incarnates. This is our Higher Self being on
the other side of our emotional life. Staying there permanently, it

allows us to enter into dialogue with more subtle spheres. This part of ourselves remains in transcendent spheres, connecting us to the World of Stars. In general, humankind, unfortunately, has forgotten its supernatural part, and hardly thinks about it.

Therefore it is a matter of course that the messengers say:

"Listen: Consider the value of knowledge!

Who lives with a consciousness of matter will never reach a higher spiritual value."

In 1995, the following message was given:

"Listen: You live on an island of probation and evolution.

There, you have to learn to divide spiritual from material values.

You have the possibility to familiarize yourself with the cosmic laws.

They show the polarity of life.

Listen: The universal laws are the connection with all lives.

All is One. Understand!"

3. "Listen: The connection is important!"

The researcher states, "True recognition can be discovered; it will be revealed to man's being connected with the Spirit Worlds."

But to access that knowledge, we must go far beyond terrestrial limits and conditions:

"Listen: You want to know many things.

Try to learn to overcome your normal awareness.

So you will experience much.

The mind in its freedom connects to galaxies."

The path into inwardness is like entering another space of knowledge, where everyone's Spirit Guide will take him by the hand and help him to move forward. All is born of Spirit. So be attentive to that which is imperishable, permitting yourself to soar to higher levels, illuminated brightly, and spending some time there. The Zentrale emphasizes the importance of meditation and inner concentration.

> **"Listen: For getting a good and constant relation to the other plane of being, an inner, spiritual preparation is of the utmost importance."**

Connecting with the Spirit World will help to learn and understand those in the spirit world, but also help us to grow spiritually, to develop and stabilize our inner self. The messengers often tell us that, when we ask them, they will help and even work together with us:

> **"Everyone, who asks for help, receives help."**

> **"Call me to you. Then I will be very close to you. I can do a lot for you if you ask."**

The invisible can sense human suffering, making it part of his life.

> **"When you are sorrowful and helpless, when pain and illness are knocking you down, think about it and ask for help."**

Although man himself turns away from spirit, invisible help will arrive at the decisive moment, when men remember. Generally, at a time of crisis or illness, we retrieve the song dozing in its depths.

> **"We can work by your prayers."**

"The other world," the researcher explains, "will always give help, even if it does not always arrive in the way man is expecting. The invisible accompany us, knowing us intimately and supporting us in every moment of life, particularly when we are in need. Everyone is able to sense this spiritual touch if he works at maintaining that union with the Other World."

Having pain is often what helps us get back to basics and become sensitive to other beings and to the life-giving source of the Spirit.

"Listen: Humans have difficulty bearing grief and misery. But it is the only way to understand the coherences of life."

Then the Zentrale continues the statement with words, deserving special attention and so providing another insight on significant realities:

"There are higher developed spirit entities who have chosen to incarnate to give the possibility to other human beings to mature because of the particular situation — for example mentally handicapped human beings. They are coming from the highest level."

Sanaedes, a Spirit entity from a very far distance from Earth teaches us:

Sanaedes: **"To know true happiness, it is important to experience even pain. Everything is accomplished in a happy human life. The soul must be able to endure valiantly even the painful and heavy hours of earthly life to become receptive to the true happiness you should all aspire to have."**

H.O.K.: "Many people here in the room would like to know what is true happiness on this earth?

Sanaedes: **"But you should know that true happiness is quite different from what you imagine. It is not wealth. Wealth can even mean misfortune; it can harm the possessor and impoverish the soul. To know true happiness, it is important also to experience the pain of life."**

Everyone, whether rich or poor, young or old, with or without culture, is capable of listening to the ineffable and wonderful message of hope that is singing through the world of tears and sorrows.

On his way of life, man never walks alone, and his step, sometimes heavy, is always accompanied by lighter periods, even if he does not notice them. The researcher once asked Sanaedes of the Fixed Star Sirius: "Do you hear me?"

Sanaedes: "I always hear you."

At another time:

Sanaedes: "I know you very well."

H.O. König: "I know you too."

Sanaedes: "I am with you everywhere."

H.O. König: "Help me to consolidate the contact. I have so many questions."

Sanaedes: "Be sure, you have help from Heaven."

This dialogue occurred between H. O. König and an invisible presence at a difficult time of his life:

H.O. König: "Can hear me? Who are you? "

Hubert: "This is Hubert."

H.O. König: "Do you still know me Hubert?"

Hubert: "Hans Otto."

H.O. König: "I feel you're always there."

Hubert: "Every hour and day."

H.O. König: "I can feel that. You know I need all my strength for this test. It is not so simple."

Hubert: "I give you all my strength."

H.O. König: "Thank you."

In other experiments with the HRS-System, HOK had this dialogue:

H.O.K: "Who's there?"

Sanaedes: "It's me."

H.O. König: "Who are you?"

Sanaedes: "Sanaedes."

H.O. König: "Do you have something to tell me?"

Sanaedes: "White Angels are standing next to you."

At another time:

H.O. König: "Who are you?"

Sanaedes: **"Sanaedes."**

H.O. König: "I greet you cordially."

Sanaedes: **"I greet you too."**

H.O. König: "I know we have a good relationship."

Sanaedes: **"I'm much more connected with you than you can imagine."**

H.O. König: "Since what time have you been linked with me?"

Sanaedes: **"I have been with you since your birth."**

H. O. König received another message one evening using the radio method: **"An Angel will call."**
The Invisible World was referring to the angels who are the messengers from the Beyond, from the Spheres of Light. They accompany us in every situation of life. Their presence can be sensed deeply by our inner self, and their whispering inspires us to become who we are.
"Is everybody accompanied by an Angel or Guardian Spirit?" someone asked in an audience.

"So realize always, that you have a soul with you,
loving you very profoundly and deeply."

Another person from the audience asked, "Sometimes I do not have the strength."

"Don't become weakened by your concerns.

Their luminous wings can touch human beings,
when you call them, when you let them establish
the bond and you allow them free rein to stay with
you; then they will give you strength."

Man plays an active role in finding his way to true happiness. No power on Earth can buy or coerce the voice of the invisible, but everyone is able to welcome it. **"Listen!"** The Zentrale does not stop saying it. Angels, who are entities of these luminous forces are often represented as having wings, the symbol of flying, the symbol of all those who are able to connect the sensitive with the supernatural, touching the soul of the one who takes the time to listen to them.

Their presence becomes a tangible reality for everyone who turns away from the outside world to give way to the World of Spirit. During communication by technical devices, they are among us. Their presence can be sensed physically in various ways, sometimes by cold or hot drafts or by a light tickling. Often a powerful light energy pervades us, causing tremors in our body and soul, giving us goosebumps. **"We will touch you or send a field of energy that lies over you. You will feel it."** At the end of a live experience with the system HRS-or-UDS system, the Invisible World has repeatedly sent energy in the form of pulsations and loud noises:

"We will send a form of energy now, for certain
people, who need this force as stabilization of their
psyches and bodies."

They remind us to proceed with our spiritual development:

"Learn to master your thoughts!"

And,

"Only your thoughts are luminous."

They make us aware of our purpose in life so that they can reach us, sustain us, and give us mental inspiration. They encourage us to discover the inner being and to travel on our own inner journey into the freedom of mind.

This path that is linked to the authentic depths of every being is unique and unparalleled. No bondage, but freedom; no servile listening, but inspiration and reflection; not by fear, but by confidence.

"Trust yourself and your inner voice!"

"Take only one path, yours, and not the many paths that will be proposed!"

"You must learn to get to know yourselves!"

Then some spirits convey that the love of people on the earth plane is of great help to them in their new lives:

"Mom, your love for me is a big help for me."

Or they communicate to us that the structure of our minds can create a force field for them: **"Thought-structure gives force field."** They emphasize strongly and express often, that they need our contacts as much as we need theirs:

"Suffering are the souls; contact helps greatly."

How is it possible that spirits, having just left their physical bodies, still need our love and our support? Their messages have taught us that after transitioning, the human being is not suddenly transported into the Power of the Divine Light. Instead, further development awaits them after the transition. They make us understand that human help is necessary for the growth of the people now invisible. Therefore, human beings still on the earth plane play central roles in the lives of those in spirit, much more than we realize.

**"Everyone must understand that here you play a
supporting role, going infinitely far beyond earthly
life."**

Through the incarnation, we can act and change things non-incarnated Spirits cannot. And our thoughts are part of a large system, where everything is interacting.

**"The universe must be penetrated and directed
by the forces of the Spirit. That is why man is a
necessity in God's eternal plan."**

Because the Invisible World decides to open or close the contact window, it is good for human beings to prepare themselves for their visits and our growth in spirit. It is essential not only for us to be aware of their presence and not oppose it, but also for us to connect ourselves with the Universe where ALL IS ONE.

**"Listen: The only human who will grow to have
good contact with us is one who takes the trouble
to understand the universal laws."**

4. "Love is life forever!"

Let us take the time and allow our thoughts to travel into the Spirit World, sending them our love and attention. They keep on repeating,

"Do not forget us!"

Another question from the audience: "How can I reach you?" This was their reply:

"Call me to you. Then I will be next to you."

The Zentrale added,

"Do not think only of your loved ones."

**"We are all a unity of life, even animals and
plants."**

And in 1991 the following message was transmitted over the Infrared System:

> **"Listen: Every person can make good contact with us if he has learned to do it. Knowledge and love are the prerequisites."**

The word "love" keeps coming in again and again. Love? Romantic love, unconditional love, divine love, compassion, passion, everyday love shared across all seasons and all ages. What is its meaning here? The Zentrale gives love a special meaning, far from strong feelings and emotional needs. They mean "com-union" "union with," or "the fact of being connected with." They do not limit this affinity and attachment to human relationships, but insist on extending it to all that is alive, including rocks, plants, and animals:

> **"Love means being in communion with all that is alive. We are all one. Recognize it!"**

> **"Recognize that we and you are one unity of all life. Plants and animals are in it."**

Love understood in this way is the most powerful force of all:

> **"Love is the highest power; it is stronger than anything in the world."**

This power of love will blossom plenteously, bearing its finest fruits, when it is experienced in communion.

> **"Listen: The power of love can only emerge by a union, a communion. Only in this way will your being flourish."**

> **"Love is life forever."**

There is no warmth or sentimentality in the words of those who live beyond our space-time universe. In contrast, they continuously emphasize how important the intensity of our ability to love deeply and dearly is. Our contact, relationships, and even the meaning of life depend only on this intensity of love.

> **"Listen: The meaning and value of life is love and the union with all that is alive."**

Our will is not what determines whether we get in touch with the Spirit World, but the efficacy of our love, knowledge, purity of heart and mind, patience, and persevering in learning about the universal laws. Their messages still say:

> **"The death day is predetermined."**

> **"The death day for all life is destined, whether by disease, war or natural disaster. Earthly man creates his own destiny."**

And also, they say,

> **"From the universal point of view, time does not exist."**

> **"One hundred years for you is for us one second."**

The Zentrale emphasizes,

> **"Listen: There is no time."**

> **"Time does not exist."**

> **You have created time for yourself."**

5. "Listen: Death is Birth to a New Life."

> **"Listen: The death, as you call it, is a necessity as the birth.**

> **Without it, we cannot live."**

Death is as necessary as birth!

> **"Listen: You live to die.**

> **Your earthly life is only a shadow of all Being/Sein.**

What you call death is important to survive and to develop in mind."

Death is birth into another form of life, where human beings actually finish their battle on Earth.

"Here the human terminates truly the earthly fight."

We receive the opportunity to continue the adventure of consciousness in the experiences of another order. The spirit, finally released from the heavy burden of the physical body, continues its journey in order to grow spiritually in more subtle and luminous spheres.

"Here comes a new life."

"With the new knowledge I can develop here."

All worlds are interrelated, but they are also distinctive. To understand the differences, it is essential to learn from our **own power of mind**. On Earth, where the material is solid in appearance, we can encounter embodied spirits on all spiritual levels. Everyone can learn to recognize them and the necessary distinctions:

"Every error in the interpretation of a human being contains an error in the interpretation of the Cosmos, because man is the microcosm in the macrocosm."

"This learning seems possible only when incarnated on Earth," H. O. König states, "because, in the Invisible World, the spirits with different levels of spiritual maturity don't meet again. They are grouped according to certain criteria that are unknown, but do actually exist, according to the Zentrale."

"Consciousness," "Spirits," or "Souls" are the experiences of a new order that allow the pursuit of spiritual evolution. The purpose of the evolution of personal consciousness seems to be part of the total union with the cosmic order, also called the divine order.

"Listen: Every human is subject to an evolution; he must cross from the basic substance to the cosmic realm.

There is no spirit being who has not crossed the interlace of existence."

The frequently asked question, then, is whether the purpose of this development is not completion of life in God. The Zentrale has used the notion of God, but almost always preceded by **"as you call it, as you say."** According to them, God is this force of love that is there for all living creatures, and that guides everyone to his Eternal self:

"Listen: God is there for all life.

"Everyone contains God within himself; it is a force of love" by which "all life is forever." And another spirit ensures:

"If God is in you and in me, we are sacred."

The teachings of the Zentrale, however, differ from the image of a personal God:

"God doesn't exist like your imagination. That is an absolute power-energy of the Universe: eternal, immutable and alterable Universe."

And about religion they said:

"The religion, you created it for yourselves."

Messages from deceased people tell us that their beliefs and inner images differ from those they held in their earthly life. The researcher recalls his contact with a very pious woman, convinced that all that was good and right came from a loving God, omniscient and omnipotent. From the beyond she informed him: **"I found God."** But someone else, who during his earthly life had doubts about the existence of God, communicated: **"I searched Gods, not found."**

But what of the messages about God transmitted by automatic writing?

"I prefer to refrain from giving any opinion on the writings received through automatic writing," H. O. König states, "because that is a different path, and difficult to issue a meaningful statement. I continue to explore with the technique. The human mind remains a great mystery. We do not know what is rooted in the depths of a being, or how his mind really works, even if the person never received religious education or he completely turned away from religion.

"In my experience, if the image and reality of God will advance the person on his way to his Inner self, the Invisible will use them to transmit certain things, whether that person is religious or not. We still know little about the functionality of our mind, about the Spiritual World and the interaction between them. But far be it from me to want to distress those who have a close relationship with God. For them, it is good and important. I fully respect their way of thinking and being. I am only passing on what I have received in such messages."

6. The World of the Beyond: "Representation Dominates!"

The Earth is **"the prelude to the symphony of real life."** How are we to imagine other worlds?

"We receive only partial images, never a complete overview," the researcher admits. "It's like asking a human being to describe our Earth. We get as many different descriptions as different people. The same pertains to descriptions of the afterlife. I found out that each entity describes his new world from a limited perspective, where it is located.

But using words like "there, where it is" assumes that space does exist. However, things are a little different. This is the point of view adjusted according to our level of consciousness. Those diverse planes are subject to different levels of development, and are therefore variously described. But even the term **"evolution level"** itself is a definition they use for us, so we can conceive of the representation."

Thus a person who has just died can never give a complete overview of the World Unseen. She is still too hampered by her way of thinking during her earthly life. She will lose that, however, in her new experiences of consciousness in the other life plane.

The answers are also related to the experimenter and the audience. When a collaborator H. O. König was asking his deceased son if his world resembled what she had read in books, he replied: **"Let it be riddles; a better time will come."** Another reply: **"You do not get to understand it."**

An entity from one of the highest cosmic levels certainly could tell us much more on the subject. But H. O. König continues: "I found that entities do not always respond immediately. For some of my questions, the answers came years later, when I was ready to understand. And they do not answer all our questions during the live experiments. They select and adapt them to the level of the audience.

They certainly do not reveal all their knowledge to everyone. They talk about the evolution of our planet, for example, only with certain people. To my question of how we should imagine the Beyond, they said: **"All that what you can imagine."** This brief response says it all! They let us imagine that we ourselves create our future life."

7. The creative power of spirit, a great responsibility

"Listen: Do not despair of things you do not understand.

You are there for learning.

Be critical in what you do!

The way you will prepare yourselves.

Keep the polarity of life in mind."

Even if the hermetic laws, in particular, were long ago, man is hardly aware of his creative power of thought both in our world and in the hereafter. The power of thought is already very great for humans incarnated on Earth.

The Zentrale said:

"Keep in your mind, that you are the originators and the creatures of your own life."

In parallel worlds, this power is immense, because everything we think instantly becomes reality.

Our reality will also initially be determined by all the important thoughts recorded during our earthly life. The Zentrale compares our consciousness to a computer that backs up all data. When we leave the physical body, we will face these data again. The man with an extraordinary creative power will bear a huge responsibility.

H. O. K: "Can you describe your afterlife?"

"Representation dominates."

"Once in the World of the Beyond, what I create in thought will immediately become real," says the researcher, "and what I think here and now will create my own future in these worlds. My way of thinking and acting on Earth therefore is certainly not meaningless, but determines my future life. We reap what we sow in this world."

"Listen: No life will end.

Every life begins with new experiences.

Prepare yourselves for your life to come.

Learning—the laws of the universe—are premises."

Believing only that in physical death the person will be transformed into a being of light by the hand of God is a willful misrepresentation. On the contrary. Upon arrival in more subtle spheres, it is not a God who will judge the person, but the conscience of that person. "We will be our own judge," the researcher says. "It is we who bear the responsibility of all our actions, past, present, and future."

"Listen: Jesus Christ did not die for your sins.

**This way of thinking is to abdicate the
responsibility."**

Human beings are originators and creators of their own world and their own existence, part of the System of cosmic order. Each person holds the key to the cosmos in his own hand, penetrated by a divine order. That is the message from those who have gone through the whole evolution.

H. O. K.: "Who judges us when we arrive on the other side?"

**"Our sin is the punishment inflicted on us by
ourselves that creates self, our own court."**

What are called, in Christian language, "hell" or "heaven" are the confrontations of our own life.

H. O. K: "Much has been said about hell. Many people ask whether it exists."

**"The self-destruction of evil is that which you call
hell."**

H. O. K: "Are we responsible for all our actions?"

**"You, the maker is, for everything he makes, is
responsible.**

Each one holds the key himself in the hand."

Question from the audience: "Will criminals and negative beings be punished?"

**"Each monstrosity, every evil act of the human
rebounds upon himself.**

**They are rubbish of the soul, as they were formed,
they disintegrate."**

**"Big sinners or criminals are taken into the World
of Shadows or they are still banned near Earth
where they will first, at the place of their wrongs,**

revive themselves all the harm they have done to other humans."

At another time, one of the audience asked, "What happens to people who commit suicide?"

"The biggest fallacy that a human being can commit is suicide. The problem, from which he wanted to run away, is not resolved after this act. He bears it as a heavy burden, carrying it around in the afterlife. The flight from life is forgiven only in special circumstances. Extreme pain or aberration."

"It would be a big mistake," the researcher insists, "to condemn these persons. We have no right. The deep, underlying reasons for taking one's own life are diverse. The being who kills himself, however, is not abandoned by the other world. He receives assistance not only from the Spiritual World, but also from people who contact the hereafter, sending them to those poor souls to help them out of their spiritual impasse. The afterlife sends out calls to all those who can hear their cries for help."

The experimenter then enters into dialogue with this entity and can give him clarity, strength, and love to empower him. Prayer and thoughts of light from relatives may also help these spiritual beings.

8. A world with different vibration planes

From birth until death, we live our lives in a plane of the universe where matter seems dense. We need a body, a temporary home for our personal consciousness, made up of different layers.

"I was trapped in my body, now I'm free."

After the great transition to other realms, experienced differently by every being, our personal consciousness still exists. It does not disappear during the separation of the body from the astral body, but it returns, a leaner body, to its homeland.

"After the earthly death we all remain in full possession of our personality, which therefore means that we also keep the memory."

And another question:

> H. O. K.: "What changes when we move to the other side?"

"We are keeping our personality."

The researcher remembers a man with cancer who had the habit of drinking whiskey every night and toasting friendship, his whiskey in hand. Shortly before his death, the man had made a special request of the researcher; they had a drink together, a sign of their unspoken bond. When the time came for him to leave, the man said in a firm voice, "Hans, to the next whiskey, together." He died two days later.

Shortly after his transition, he got in touch with H.O.K., saying his name and **"Connected, Hans."** And then humorously, **"Time didn't go around for another whiskey."** Five years later, this man came through again several times, but in quite another way. He gave information about the new tasks he performed in the other realm, like helping souls arriving there, completely lost and not knowing what was happening to them.

Question from the audience: "Can our soul, in some way, be destroyed?"

"It is also called the astral body. This body is not to damage, to injure or to destroy by any earthly force."

This spiritual body continues to live, as expressed in the following message:

"That's funny. Have two bodies, one which was buried, the other one the same, but healthy."

With this astral body, we can continue to feel:

"Yes, in your soul-body you have a soul again,
therefore a keen inmost being with which you feel
and sense again."

At another time a voice said:

"I have here new senses."

All the senses of human beings are much more refined in spiritual dimensions. In these realms, we take along not only our earthly experiences, but also our personal needs, even if these are reduced, little by little, by new experiences:

"Every human first takes his needs along there,
reducing themselves over time."

The researcher explains, "Every soul arrives at the other world at the first level, as they call it. "Many are sleeping" on that first level. "Others believe to go on living as they did during the earthly life." Some are found in "hospitals," an expression the Invisible precedes with "as you say," a place where the spirit has the ability to regenerate. From there, a sorting takes place.

Each consciousness pursues its own path to other levels of existence, as they call them so we can understand. The level in which we will live is assigned according to criteria related to the maturation of spirit and the spiritual knowledge of each entity. Entities having the same level of consciousness meet and live in union. Apart from three negative levels that exist, mention is made of seven levels of evolution, which differ clearly from one another. The Zentrale described them as follows:

"Listen: You know there are seven planes in the astral spiritual domain.

These planes differ from the color representations, as you say.

First plane: blue.

Second plane: green.

Third plane: yellow.

Fourth plane: orange.

Fifth plane: red.

Sixth plane: purple.

Seventh plane: white.

The seventh plane is the highest cosmic level.

Listen: The cosmic area is the highest area which a spiritual being can reach.

It contains all vibration of the planes."

"Our interlocutors from the spirit world have tried to illustrate using the color spectrum, which gives us information about the diversity of frequencies. Nevertheless, their reality remains incomprehensible and fairly impenetrable, even if we can be sure that different levels of consciousness and evolution definitely exist. But a concrete image of the system and its operation remains for us unattainable."

"Listen: There are seven planes of development in the spiritual astral area.

They differ in the strongest way."

"Listen: There are seven positive planes and three negative planes, whereas one incarnation only will be difficult without knowledge."

"Listen: The knowledge will be decisive, in what life plane you will be incarnate."

"Listen: There are different development planes."

"You know, there are seven planes of evolution."

"Every development plane means a new birth, that you call death, but that has nothing to do with your earthly death."

In these other dimensions, as the Invisible states, we communicate telepathically:

> **"Over here there are no language difficulties.**
>
> **We are able to communicate by telepathic transmission."**
>
> **"Over here there is no popular or racial segregation.**
>
> **We are all the same.**
>
> **There are even no language or communication difficulties.**
>
> **Likewise there is no religion."**

And they often insist:

> **"A beautiful, supernatural world it is, in what we are living.**
>
> **There is no pain, neither for men, nor for animals.**
>
> **But probably it is too difficult for you, to understand this."**

Faced with the reality of the subtle worlds, we are left without words, in silence.

> **"Live your life, do not try to live our life."**

9. "Listen: Incarnation is not bound to your planet."

Are there souls who one day choose to leave the subtler worlds to reincarnate in the densest matter, the Earth? Is reincarnation real? "We still know relatively little about reincarnation. And as it's my desire to hand down knowledge grounded in my scientific approach, beyond the realm of faith, I don't care to elaborate on this subject.

From the perspective where **"Each human is subject to an evolution, he had to make beginning from the basic substance till** to the Cosmic Realm," it seems logical that reincarnation does exist. Many messages form the Beyond have confirmed it for me. But contrary to certain beliefs, the human soul does not migrate from one body to another, for example from that of a human to that of an insect." The researcher takes out pages of dialogue when he questioned his invisible friends on reincarnation. Here, for example, is an extract of a recording made in Kaarst, a city in northern Germany, during a convention in 2002.

> H. O. K: "The conference theme is reincarnation. Can
> you tell me something about it?"

"Loads of germs of human souls are still lying in the garden of expectation."

Question from the audience: "Do we, in the next incarnation, come together again with the persons here on Earth with whom we were connected?"

> **"One day in this earthly life or as mental sin, purification, and in another incarnation, he will meet that person again."**

Question from the audience: "Will we have the same problems or will everything change?"

> **"And then, in some cases there are problems again and difficulties again."**

> H. O. K: "I still have so many questions to ask about
> your messages."

"Here shall not and cannot be answered to all statements."

Then the researcher read out the messages received by radio recording. One evening when he called a certain person, he was told:

"She will be embodied and not return."

Or the following dialogue :

H. O. K: "Why did X not get in touch again?

"Her voice is taken."

H.O.K.: "How should one represent reincarnation?"

"The dead will here always die."

H.O.K.: "Can you tell me why he wants to incarnate again?"

"A man, like he would be again".

H.O.K.: When did X embody?"

"He died here in August."

And the following:

H.O.K: "You said you would embody again."

"On the other side."

H.O.K.: "And then when?"

"One day through the door."

H.O.K.: What changes in the incarnation?"

"Gender changes, just as the breed."

H.O.K.: "Do you remember a previous life?"

"Time is only a sentence until man dies again."

What we call original sin is a legacy of reincarnation. But this is subject to laws too difficult for us to grasp:

"There are secrets about rebirth, rules, regularities, maturation- and stabilization-processes of human substance in the course of karmic involvements," as Sanaedes said.

Even the Zentrale transmitted us a wonderful message of hope:

"Listen: Every incarnation is voluntary.

It is a facet and fulfillment for evolution.

The regularities of evolution are given.

Choose you your parents by yourselves!

The parents give you the needful, to get through this incarnation.

Everyone during his incarnation is aware of his situation."

This implies that Incarnation is not mandatory and not limited only to our planet:

"Your incarnation is not related to your planet."

There are many stars and planets where it is possible to develop in mind and spirit. As Sanaedes answered the day the researcher posed the following question: "What are the tasks of human beings on Earth?"

"The All is constantly expanding, new suns and planets will be born, and everywhere there are tasks to be fulfilled, by you human beings, who here are born, grown and matured, on this so small, but so incredibly important Earth plane."

And the Zentrale's specified:

"Listen: The earthly existence is limited.

Incarnations on other planetary-systems are possible."

"Why is man born here on Earth, and not on another star?" the researcher asked another day. Sanaedes responded:

"You, the human beings, you were born on this little Earth plane, because you have to continue living in the most beautiful realms of light where you can get to grow spirit and soul."

H.O.K.: "On Earth we have developed a lot in terms of technology and science, as in medicine. Does this development also exist on other planets?"

"Until now, scientific progress did not lead humanity to unity, peace and well-being, health and true happiness."

10. Man is not the crown of creation

The Earth is one of many planets in the Universe where life exists. And on this Earth plane, man is a living creature among other living creatures, such as animals, plants, and minerals. The first time the Invisible World spoke on animals and plants was in autumn 1987 during a conference in the city of Büdingen, Germany. The Zentrale spoke about the worlds in the hereafter, where there is no gender difference or separation between races or nations, or problems with language or communication or religion.

Then they added:

"We all get to communicate with plants and animals. You will appreciate this, if you learn it."

"Listen: Your attitude towards animals will have karmic consequences. You will have to bear the blame."

In May 1989 in Büdingen, the Zentrale transmitted by the Infrared-System:

"Listen: Recognize that we and you, we are one, one unity of all life, plants and animals inclusive. Without them, you cannot exist on Earth."

Then she continued with a sentence that shocked many conference participants:

"Listen: If you believe that man is the crown of creation, then you are wrong. A plant or animal can be
entities, just more developed than you."

In 2015 this was received by the UDS-system:

"The plant, the animal, the human, each crystal has
a soul."

Between 1987 and 2015, the Invisible regularly drew the attention
of humans to the world of animals, plants, and minerals.

"Listen: The meaning and value of life are love and
communion with all life forms.

Listen: Everything is giving and receiving.

The plants and animals give you love and nourish-
ment.

You should give them the same.

But all that you make is taking."

After their death, all animals and plants immediately join
the third level where they continue to live in a world of love. The
Zentrale affirmed:

"Many of them will recover, from what was done to
them during their reincarnation.

This is a great right."

One day, while the Königs were making contact trials with
their radio in the woods, near a huge stone they heard the following
message:

"Failed man.

They ignore the laws of nature.

Traumatic legacy."

Then again:

"For so many, many years, I give my strength.

They call me the health of the woods.

All are passing me by."

As already noted in this book, crystals are beings of light, and of extreme significance for the contact bridge.

"Whoever having a crystal, treat him like a brother.

It supports you in any situation."

No crystal resembles or resonates with any human being. If we have the chance to meet an inhabitant of the mineral realm, resonating with our Self, we must nourish and cherish it. It will help develop the level of spirit. Its energy can facilitate the opening of doors to the Invisible. **"Crystals are the keys to the connection."**

They are living beings that can transfer a tremendous force on man, if he connects with them spiritually and begins to listen.

Through contact trials made with the HRS-System, the following dialogue occurred between H. O. König and Sanaedes:

H.O.K.: "What is the position of man in the universe?"

Sanaedes: "According to the cosmic laws, man is the microcosm in the macrocosm."

H.O.K.: "Man believes that he is the crown of the universe, occupying a special position on the Earth."

Sanaedes: "Anyone who becomes aware of the cosmic realities, will recognize that the microcosm, man lives not only lives restrictedly, but in a dungeon corresponding to his own desire and to the world of his own needs."

H.O.K.: "You already said a lot about animals. What about their incarnation?"

Sanaedes: "The living areas of the animals, even the one of the animals incarnated in an earthly body, correspond to their current level

**of consciousness, which already will carry on
the following evolution stages."**

H.O.K.: "What must we do to change your point of
view?

**Sanaedes: "It is very, very sad that you have al-
ready rendered extinct a lot of animals of different
species by your greed for wealth and luxury."**

The Invisible invites humans to relearn communication
with everything that is alive. We are capable of that, because everything alive has a soul and can understand the language, even if
not always through words. In the other realms, all communication
proceeds telepathically. Man must remember this innate path.

"It (the communication) begins resurfacing in itself if we
take the trouble and time to connect with all that surrounds us: humans, animals, plants, or minerals. Then we will realize that all life
has a consciousness correlated with its level of development and its
situation in life. We claim to believe that man is more developed
only because he has other intellectual abilities. How sad to see the
behavior of humans toward nature that surrounds them."

11. A Warning and a Message of Hope

Just as a spiritual being, living millions of light years away
from our planet, so effectively expressed:

**"Too much poison flows from the rivers into your
seas. This causes the disturbance of nature, resulting in terrible consequences for you. "**

**"Too much poison finds its way into the atmos-
phere. You know it and the consequences, and you
make no move to do something. "**

These messages came from Sanaedes, but she is not the only
one who tells us about the great concern the World of Stars who are
watching over our planet share. The Zentrale also informed us,.

"Listen, what you call progress is the preparation for the doom of all life."

One day, H.O. König used the HRS system to ask how the Spirit World sees our actions on Earth. They answered,

"Man in his arrogant selfishness destroys everything until he can control it."

H.O.K.: "What can we do?"

It has been repeated on several occasions that the Spirit World is sending signs, and they are constantly emphasizing the importance of sharing the results of research with humankind. Experience has shown that it will be understood by very few people.

"Great souls grasp the Great,
Low souls reject it.
Understand it and be happy about it!"

At another time, when Sanaedes told him that the emptiness of the human heart is becoming more and more extensive on our Earth plane, the researcher said: "If what you say is true, we could give up on everything."

Sanaedes: **"Never indulge in despair!**
Very powerful, spiritual forces are protecting you."

In 2015, we received messages full of hope for the whole of humanity:

"It breaks at the turn of time. Be joyful for the
future, because great rays will be guided from
the cosmic suns down to the Earth. We are
coming from your future.

Do not despair; there is help!"

H. O. König concluded, "Seeing the light, the goodness, and the beauty without shutting out harsh realities, being messengers of hope without losing ourselves in illusion, is a paradox we live with every day. It is a beautiful, important task for humankind.

**"Listen: If you want to change the World, change
your own world, the world in you! Spirit overcomes matter!"**

Even if our society is convinced of the contrary."

**"We know the time is ripe for an expansion
of consciousness."**

Epilogue

"Listen: Many believe, but do not know, that there is no death.

There is only life, in affiliation/togetherness, forever.

Everything is transformation.

We are reiterating it again and again: We are alive; we can see you and talk."

The Spirit World draws our attention to the difference between knowing and believing. It is important for the human being not only to believe, but to know that all is born of spirit and that physical death does not mean the end of life. Today, however, collective knowledge is determined by the discourse of science, the official authority to which most people refer.

"From the research of Isaac Newton and the mechanical era," the researcher says, "what science can prove is considered true." This means that every phenomenon in nature must be quantifiable, qualifiable, and reproducible under certain laboratory conditions. Now, it is clear that the apparitions of metaphysics are exempt from this exposition of proof, and therefore they are relegated to the fields of religion or spirituality. The consequences of this way of thinking are manifest in our consumer society with its ecological, economic, and human disasters.

"In physics, as in science in general, there are no permanent truths or proofs. There is only a sequence of probabilities, conceptions of truth, and indices. In forty years of research, I have gathered enough clues to demonstrate the permanence of individual consciousness beyond death."

Do you believe that one day science will include in its field the results of your research, like those of other notable researchers, even if they overturn existing paradigms?

"New paradigms in science can be found and established only by the emergence of a new way of thinking. History tells us that this process takes a long time. I agree with Sanaedes, who said, **"Believe in the immortality of the soul, scientifically unsustainable."** For now, my research is not and cannot be upheld because it requires a fundamentally new way of seeing the world.

I feel a bit like Don Quixote fighting with windmills, declaring a way of thinking already obsolete but still celebrated. The perspective should be reversed, where we accept that the mind exerts an influence on matter. It is to be hoped that with the evolution of science, one day we recognize that other dimensions do exist and interact with our physical world.

From that moment, science will begin a thorough examination of the phenomena on which it now casts a contemptuous glance and will move on to metaphysics or transcendental physics. Then we will define a "new field" of which "Being in itself" may not be more well known than "Being in itself" of the electromagnetic field, but unlike the other could already be calculated and used for practical purposes.

"I am convinced that we are at the heart of a process of significant change, a paradigm shift in consciousness and thought. In this change, quantum physics could be an important pillar. The goal is an expansion of our consciousness over the spirit of nature and matter. It is in our evolution to abolish the boundaries that exist between different levels of consciousness. But I am only a tiny grain of sand on planet Earth, so . . ."

The story is not over here. So many silver threads weave the Universe. Notes resound from the infinite, awaiting attentive ears. We are surrounded by mystery and live within the miracle of life. In every being there is an ancient song that whispers and calls, inviting us to discover its truth and love. A melody sparkles with hope, despite absurdity, nonsense, death, pain, disease, harshness, destruction, cruelty, and human injustice. The Spirit breathes and lives in us.

Sometimes you can hear its whispers, sense its scent, or perceive one of its many forms. Hans Otto König listened to its messages and traveled a unique path without being influenced by the sounds and patterns of the outside world. The Spirit World gradually granted him access to its realms and made him understand that we can not only reach the approach of love, but also the approach of knowledge, which is reality. If his research allowed him to capture thousands of voices tearing the veil between the worlds called "here below" and "beyond," it also made him feel **humility for that mystery, that is behind All that is Visible.**
In his laboratory, in all modesty, but in a soft but firm voice:

> **"Listen: Help, please, to transmit to all human beings that life after death exists. Believe me, it is important to know."**

May the wind take these few notes of light into the homes of all beings who wish to hear them, gently pour balm over their sorrows, fertilize their hearts, thoughts, and souls by illuminating them with the power of knowledge that is love.

> **"Listen: The meaning and value of life is the love and affinity to all what is life."**

> **"Love is life forever."**

Anna Maria Wauters, Antwerp, 2015

Thanks

Many thanks to Jacqueline Kelen for her invaluable help and support in writing this book.

With gratitude to Mia and Alfred Wauters-van Engeland. I say a big thank you and embrace them from the bottom of my heart.

Thanks to Evelyn Meuren and R.Craig Hogan for their immense work.

Contact:

For more information on the research of Hans Otto König, seminars and lectures, single sessions, personal coaching and participation of live recordings with the greater apparatuses, please contact us:

Email: annahanskoenig@gmail.com
hansottokoenig@gmail.com

Website: www.hansottokoenig.de

Appendix

The Future for Afterlife Communication

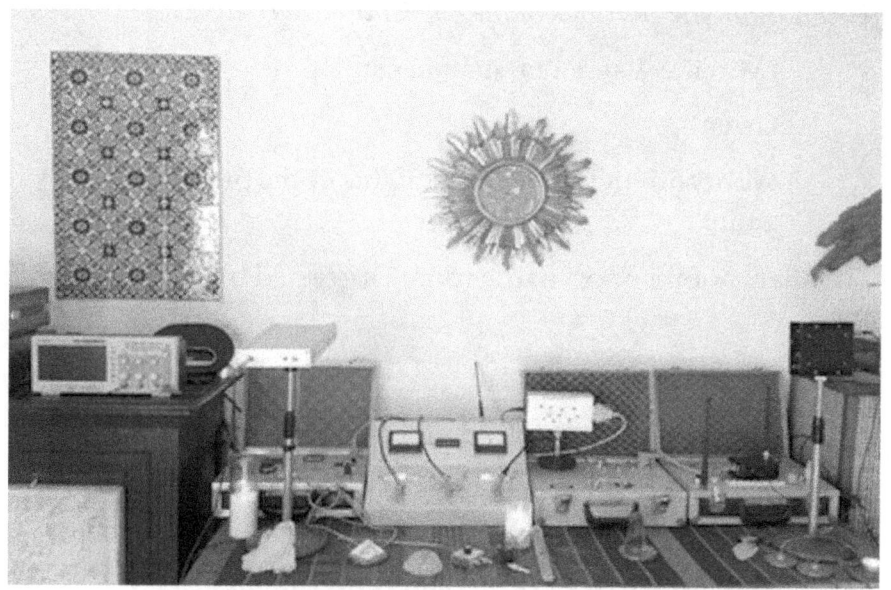

The modified Infrared System, April 8, 2017, Antwerp

Recording With the Modified Infrared System, April 8, 2017, Antwerp - Belgium.

(The received messages are in German, Latin and French.)

"We'll try to complete contact panel to Anna Maria Wauters and Hans König.

Contact field closed to Anna Maria Wauters and Hans König!"

Mireille, one of the hosts receives a personal message from "Padre Pio", having been in contact with him for several times. Padre Pio announces Mireille's name, his name and then says, in Latin, the famous two words: "sursum corda = Hearts lifted", the opening dialogue to the *Preface of the Eucharistic Prayer or Anaphora in the liturgies of the Christian Church.*

"Mireille, Padre Pio, sursum corda.

Listen!

We absorb the power of thoughts of the present group."

Christian, another host in the room, is greeted by his spirit guide Gilles.

"Gilles greets Christian.

The contacts continue to become closer.

If these are speaking in this manner, what do the ones on the Higher Planes say?

You bring new horizons and furtive thoughts.

They come well, once the connection is established, thanks to the exact frequency of the communicating entity.

Listen!

We are connected with your thoughts and we can hear your thoughts. We're perceiving them!

Listen!

We'll, thank the present group for its participation and the...... each path.

We are following and accompanying that path.

Listen!

We see a new way in the research of Hans König.

Listen!

We thank Anna Maria Wauters and Hans König for their spiritual sustainment!

We close contact! Energy field begins to dissolve!

We send you all an energy field and a field of power!"

Hans Otto König and Anna Maria Wauters during a conference

Anna Maria Wauters

Seminar at Paris -France, 2017

Hans Otto König

2017 Lab

"A RESEARCHER STILL IN HIS LAB!"

Hans Otto König, 2017